Earth at the Crossroads: Understanding the Ecology of a Changing Planet
Part I

Professor Eric G. Strauss

THE TEACHING COMPANY ®

PUBLISHED BY:

THE TEACHING COMPANY
4840 Westfields Boulevard, Suite 500
Chantilly, Virginia 20151-2299
1-800-TEACH-12
Fax—703-378-3819
www.teach12.com

Copyright © The Teaching Company, 2009

Printed in the United States of America

This book is in copyright. All rights reserved.

Without limiting the rights under copyright reserved above,
no part of this publication may be reproduced, stored in
or introduced into a retrieval system, or transmitted,
in any form, or by any means
(electronic, mechanical, photocopying, recording, or otherwise),
without the prior written permission of
The Teaching Company.

ISBN 1-59803-588-6

Eric G. Strauss, Ph.D.

Research Associate Professor of Biology, Boston College

Dr. Eric G. Strauss is a Research Associate Professor of Biology at Boston College, where he also serves as the director of the Environmental Studies Program and as a science advisor for the Urban Ecology Institute. He earned his B.S. from Emerson College and his Ph.D. from Tufts University. His research focuses on behavioral ecology and science education and is conducted in 2 primary locations—the Boston metropolitan area and the Sandy Neck Barrier Beach Complex on Cape Cod, Massachusetts.

In his research on behavioral ecology, Dr. Strauss uses methods such as mark-recapture studies, direct observation, radiotelemetry, and bioacoustics to elucidate the factors that influence the reproductive success and life history patterns of coyotes, diamondback terrapins, and other social vertebrates. These long-term research efforts are strengthened by collaborations with the Urban Ecology Institute, the Lynch School of Education, and the Urban Ecology Collaborative. With help from these partners, Dr. Strauss works to promote the stewardship of healthy urban ecosystems by improving our knowledge of urban ecological function and by engaging communities in the process of urban restoration and transformation.

Through his work in science education, Dr. Strauss strives to create authentic connections between the scholarship of original research and the teaching of high school and college science. His research studies serve as curricula for courses taught at Boston College, its partner universities, and collaborating high schools. To this end, Dr. Strauss tailors his methodologies to incorporate student scientists of all ages into as many phases of his work as possible. His main courses include Foundations of Urban Ecology, Ecology of a Dynamic Planet, Animal Behavior, and Organisms and Populations.

Dr. Strauss has published extensively, both in behavioral ecology and in science education. He is the editor in chief of the Web-based journal *Cities and the Environment* and is senior author of a nationally distributed science textbook, *Biology: The Web of Life*, and an upcoming textbook on urban ecology.

In addition to his research, teaching, and publishing, Dr. Strauss has served on the Barnstable Conservation Commission and on the boards of directors for the Cape Cod Museum of Natural History and the Association for the Preservation of Cape Cod. His research projects are supported in part by the National Science Foundation, the Henry David Thoreau Foundation, the U.S. Forest Service, the Boston College Intersections Program, and other funding agencies.

Table of Contents
Earth at the Crossroads: Understanding the Ecology of a Changing Planet
Part I

Professor Biography		i
Course Scope		1
Lecture One	An Ecological Diagnosis of the Living Earth	4
Lecture Two	Humanity and the Tragedy of the Commons	19
Lecture Three	Ecology—Natural History to Holistic Science	33
Lecture Four	Ecology as a System—Presses and Pulses	48
Lecture Five	Climate and Habitat—Twin Ecological Crises	62
Lecture Six	Human Society as Ecological Driver	75
Lecture Seven	Movement of Energy through Living Systems	88
Lecture Eight	Humans as Energy Consumers	101
Lecture Nine	Nutrient Cycling in Ecosystems	114
Lecture Ten	The Challenges of Waste and Disposal	128
Lecture Eleven	The Water Cycle and Climate	141
Lecture Twelve	Human Water Use and Climate Change	154
Timeline		167
Glossary		171
Biographical Notes		175
Bibliography		179

Earth at the Crossroads: Understanding the Ecology of a Changing Planet

Scope:

The science of ecology, whose name derives from the Greek *oikos* ("house"), has contributed significantly to our understanding of the structure of living communities and how they evolve. In this course, we will span scales from molecules to entire ecosystems, discovering how the field of ecology has developed holistic models that we can use to understand, protect, and benefit from our environment.

With so much concern and emotion being focused on climate change and the impact of humans on our planet, the science of ecology gives us a rational window through which these issues can be viewed and considered. In this course, we will investigate theories and data from models of how communities of plants and animals interact. Drawing on findings from the historical roots of ecology to the most recent work, our topics include food webs, predator-prey dynamics, biodiversity, coevolution, and urban ecology.

To understand ecology is to reveal the nature of ourselves and our relationship with the species around us. Indeed, ecology has become far more than the study of pristine nature that many people expect. Newer approaches often investigate ecology's effect on humans, as well as effects of humans on ecological systems. Through agriculture, technology, and urbanization, humans have successfully inhabited all of the world's ecosystems and now have a profound influence on ecological processes at all spatial scales. We are living at increasing densities and are creating urban megacities that contain 25 million people or more, changing the nature of global ecosystems.

The lectures in this course are often paired, with basic ecological principles discussed in the first and the human role and implications addressed in the second. Lecture One, for example, surveys overall issues for the biosphere, while Lecture Two focuses more directly on humans as causative agents and the so-called tragedy of the commons.

Habitat changes are perhaps a less familiar topic than the climate challenge. However, habitat destruction is the largest threat to our world's biodiversity, which is why habitat is a central focus throughout this course. Habitat destruction encompasses a range of issues such as deforestation, habitat fragmentation, and even urban sprawl that are a huge threat to our ecosystems.

Ecology is not a religion or a philosophy but rather a science, rooted in history and driven by data, that can create reliable scenarios of how Earth's living systems interact and respond to change. The great capacity of ecology to serve as a guidepost for the challenges of a changing world derives from 2 central tenets.

First, ecology is highly interdisciplinary and draws on theoretical sources and tools from a wide variety of research fields, including biology, earth science, sociology, history, economics, and philosophy. Moreover, as the study of ecology has matured, in-house theories of change have also emerged, such as the ecosystem services model within the National Science Foundation's Integrated Science for Society and the Environment framework. Using this theory, we can study ecology as a collection of forces that sculpt a changing landscape. Some of these changes (called presses) occur over very long periods of time, such as climate and natural selection; these changes are interspersed with short but intense agents of change (called pulses), such as storms and human-caused habitat destruction that affect the function of ecosystems.

Second, underlying the power of ecology is that engagement with natural history, especially with that of humans, is critical to understanding ecosystems. This is a novel but necessary feature of ecology as a science: We cannot understand the current distribution of species and their habitats within an ecosystem without investigating prior conditions. Ecological systems cannot be studied in an isolated experiment, unlike topics in chemistry or physics. Because ecology is the study of interactions, outcomes vary widely according to the preceding events. This idea is particularly cogent when we discuss community composition, biodiversity, and disease. Engagement with natural history and physical place makes the science of ecology a particularly powerful tool for understanding the human condition and for charting sustainable directions into the future.

The majority of the lectures in this course are divided into 2 broad sections. In Lectures Five through Eighteen, we will investigate specific forces and drivers that cause ecosystems to change over time, including the flows of energy, materials (including waste materials), water, food, and toxic substances. The other section of the course (Lectures Nineteen through Thirty-Two) focuses on ecological interactions and processes. These include population growth, migration, disease, microevolutionary changes within a given species, and the coevolution of complex systems such as zoonotic disease.

We will conclude with 4 lectures (Lectures Thirty-Three through Thirty-Six) on human responses to environmental challenges that we face as a species, with a focus on positive solutions and outcomes. Here are the most vital messages of this course. These lectures consider current and prospective contributions from science, new designs and technologies, hopeful directions for cities, and new ways of thinking and living that maintain and restore healthy ecosystems.

Throughout this course, we will engage the great message of ecology—that it is really the science of our lives. The topics will grab your attention because they are windows into the nature of the life around us. As we will discuss, the lessons of nature often go unheeded, and we have paid the price as a species, with the tragedies of droughts, pandemics, and scarred landscapes as testimony to human ignorance. However, the tools of ecology can also do enormous good, and we will learn, for example, how ecological thinking has helped humans develop the integrated pest-management tools and agricultural techniques that ushered in the green revolution. Ultimately, we are all richer for the legacy of ecologists, and the lessons of ecology can greatly inform our technology, agriculture, medicine, and society.

Lecture One
An Ecological Diagnosis of the Living Earth

Scope:

This course will investigate the wide scope of systems within Earth's biosphere and the ecological changes shaping life on Earth. We will look at how ecology helps us make sense of the past, understand the present, and investigate the possible outcomes of the future. We will focus on 2 signs that Earth's ecosystem is possibly in decline—climate change and habitat change.

Outline

I. Welcome to a course that serves as an ecological analysis of our dynamic planet.
   A. In this course, we are interested in the biosphere: the thin realm in which life exists.
   B. There are 3 core emphases in this course: humans and other organisms, boundary changes, and the human outcomes related to these changes.

II. The study of ecology is the examination of the interactions between the biotic (living) things and the abiotic (nonliving) things.
   A. There is a bit of a false dichotomy between science and the humanities, because ecology is very much linked to natural history and the ideas of narrative and legacy.
   B. The model of gathering ecological information in pristine and remote ecosystems has led to some extraordinary understandings of nature.
   C. Newer approaches have driven us to ask more integrated questions and to investigate the impact of the ecology on humans, as well as the impacts of humans on ecological systems.

III. Ecology is about what sorts of changes are possible, which are not possible, and which would be possible at too high a cost. Most of it is an amazing testament to human ingenuity and a really terrific success story.
   A. The homogenization of food quality means we can eat just about anywhere now and not get sick.

- **B.** Lead emissions in the United States have dropped almost 98% since 1970.
- **C.** Temperate forests have seen an incredible rebound from the previous century.

IV. Ecology is a science of how nature works; this course will not be an exercise in preaching.
- **A.** Ecology is a way to understand the past, to understand the present, and to investigate the potential outcomes of the future.
- **B.** Learning about ecology does not mean you have to give up your selfish best interest. Remember, life depends on the fact that every living organism will invest in its own survival and perpetuation of its genes.

V. Ecology is looking for mutually supportive long-term benefits.
- **A.** For instance, if the physical and social benefits of planting trees make a city more livable and sustainable, then the investment is sensible.
- **B.** These types of changes are in our selfish best interest, especially if the ecological transformation employs people in green jobs and the metrics of public health improve.

VI. Ecology has multiple subfields: population ecology, community ecology, behavioral ecology, systems ecology, and a new branch called urban ecology.
- **A.** Each of these plays an important role in addressing the interrelated challenges that suggest that the living Earth faces a crisis.
- **B.** We will also recognize and use the contributions of allied fields, such as public health, earth science, economics, engineering, history, and the humanities.

VII. There are 2 sorts of changes that suggest that Earth's ecosystem as a whole seems to be in decline.
- **A.** Climate changes are seen in the sea level's rise, an increase in greenhouse gases, and harsher and more severe weather patterns.
- **B.** Habitat destruction is the largest threat to our world's biodiversity, which is why habitat is a central focus throughout this course.

Suggested Reading:

Hage, *An Entangled Bank*.

Worldwatch Institute, *The State of the World*.

Questions to Consider:

1. How does the science of ecology differ from natural history studies?
2. What makes the methodology of ecology unique compared with other natural sciences such as chemistry and physics?

Lecture One—Transcript
An Ecological Diagnosis of the Living Earth

Welcome to a course that serves as an ecological analysis of our dynamic planet. Hello, my name is Eric Strauss. In this course, we are interested in the Earth. Actually, we're interested in the part that's confined to the very thin film on the surface that we call "the biosphere." Life's realm which is only about 10–20 kilometers thick. Some migrating birds can be found flying high over mountains at about 8000 meters, and some fish swim at depths as large, but it still represents a very tiny portion of the Earth that is actually hospitable to life. As the late astronomer, Carl Sagan, pointed out, it's about the same thickness as that of varnish painted on the surface of a basketball.

However, within this thin film of life resides a million named species of organisms, some with such complex interaction that it seems almost like science fiction. It is in these very interrelations that we are investigating in this course—interactions among organisms and the Earth they inhabit.

Now, there are three core emphases in this course. One is that we're going to be looking at humans and other organisms. Two, we're going to be looking at what we call "boundary changes" that have been created by humans. And three, from an ecological perspective, the human outcomes related to these changes.

We engage a wide range of topics, from trees that sequester carbon, to water fleas that clone themselves. But we also never step very far from the perspective of, "what's in it for me?" After all, living systems are so resilient because that is the central concern of every living molecule and every gene that is expressed in nature.

Our primary focus in this course is ecology as a science. This project builds from courses that I already teach or have taught, including Ecology of a Dynamic Planet, Animal Behavior, Foundations of Urban Ecology, Introductory Biology for Bio-majors, plus freshman mentoring and seminars in Ecology. And I'm very fortunate to teach with a cadre of exceptional scholars, including a physicist, an environmental lawyer, and my former high school Ecology teacher who, upon getting his Ph.D. has joined us on the faculty and runs our field station.

In my teaching and mentoring of students, I make use of my own everyday field work and experimentation, and mostly of those committed scientists that inhabit our discipline. My Ph.D. was on studying the behavioral ecology of piping plovers. They're a controversial species. You may have heard about the land closures around their protection and seen bumper stickers like "Plovers Taste Like Chicken." And unfortunately, or fortunately, depending on your perspective, some of the data that I gathered contributed to the insight that protects this species now.

This course is going to reference many current studies drawn from peer review journals. Since the late 1990s, when my first textbook was being published, people would hear I teach Ecology and Biology at a Jesuit University, and one of the first questions they have is, "How do I handle questions about global warming, or about evolution?" My response is that my approach is shaped by the limited, but powerful, methods of science.

Now, in science we certainly believe in gravity, but we call it a theory. That theory is not best tested by jumping off of a 5th-story roof, but we call it a theory because something better, as a descriptive tool, may come along, with better evidence and arguments, and we'd better be in a position to accept that as scientists.

Now, global warming and evolution are the most powerful tools that we have in our toolbox of science to explain the phenomena that we see. Both are backed by such a mountain of repeatable and peer reviewed research. However, as scientists, we must be prepared in our professional and psychological domains to abandon these models if better ones come along.

So, this is very much a science course. But I also point out that there is a bit of a false dichotomy between science and the humanities because ecology is very much linked to natural history and the ideas of narrative and legacy. What happened before is critical to the nature of ecological revelation, and that differentiates it a bit from some of the other sciences that are more reductionist. Theoretically, a physics experiment conducted anywhere in the universe, if the variables are the same, should come out the same. However, ecological investigations are embedded in knowledge that we gather about the forces that drove the situation and where it might move over time. And so in that sense ecology is a bit holistic.

Now, there is an impression that ecology is the activity that happens when you only go to distant locations. You get on a plane, and from that plane you take other forms of transportation and, ultimately, you're dropped off in some remote area, and that's where ecology begins. That's where you start to gather data. Now, in fact, this model of gathering ecological information in pristine and remote ecosystems has led to some rather extraordinary understandings of nature. For example, the great systems work of Eugene Odum in the 1930s has led to an incredible understanding that ecosystems are complex and interconnected by both physical and biological drivers.

And yet ecology also turns out to be far more than the study of pristine nature that many people expect when we say the word "ecology." Newer approaches, which, by the way, are funded by the agencies that drive our work—The National Science Foundation, the U.S. Department of Agriculture, Forest Service, environmental protection agencies, and others—have really driven us to ask more integrated questions and conduct studies that investigate the ecological impact on humans, as well as impacts by humans on ecological systems.

So maybe we should begin with a definition. The study of ecology is the examination of the interactions between the biotic, living things, and the abiotic, nonliving things.

There's a joke about ecologists (often true) that we tend to eat alone because whenever we come to somebody's house for dinner, our environmentalist roots emerge. Folks around us are in fear that we're going to be asking whether the table comes from sustainable wood, and worrying about whether they're serving us free-range chickens. But you know, in fact, ecology and a sustainable world view, at least in my perspective, isn't about giving up meat or giving up one's car. It's instead a little bit like a marriage, a long marriage, where one develops a sense of what sorts of changes are possible, what are not possible, or what would be possible at too high a cost. In fact, I think, like a good marriage, much of what we see around us is an amazing testament to human ingenuity and positive belief, and really is a terrific success story.

Think about this. For instance, we can eat in virtually any restaurant, even the cheapest fast food joint that's not particularly healthy food, and we virtually never get sick. We take that for granted. But in the not-so-distant past, people very often got sick from the food they ate, even when they were eating so-called "natural foods." So one of the great testaments of success of our human species, at least in developed nations, has been the homogenization of food quality.

Even in more traditional ecological terms, we have terrific success stories. For example, the Environmental Protection Agency's reduction in lead emissions in the United States has dropped lead emissions almost 98% from what they used to be, even in 1970. Temperate forests have seen an incredible rebound from the decimation observed in the previous century. In short, ecology can be, and I hope this course will be, a manual scholarship and a source of inspiration.

For instance, one of the things that motivates me as an ecologist is the large pool of under-served human communities, and how we tend to under-use nature as a tool for human improvement. This is a tremendous opportunity that we're just beginning to unlock as a function of the recognition of nature of humans and nature by humans. So nature has profound effects on our physical being and our psyche. We'll look at research by folks like Robert Ulrich, who found that even something like the duration of a hospital patient's stay after surgery is shorter, and the amount of pain medication they need is reduced, if these post-operative patients can see trees and greenery outside their window. Imagine something as simple as that connection to nature having physical and psychological impacts on healing from illness.

Now, it makes sense, actually. We were adapted over a very long period of time to see what trees have to offer. And it makes sense—this isn't some sort of voodoo—that solutions we've adapted for very long periods of time turn out to be beneficial to us within our own individual lifetimes.

Now, what ecology is not, and to the best of my ability, ecology will not be an exercise in preaching. It is a science that is an offering of how nature works. Here is what we know, and how we know it, and what kinds of decisions we are facing, and what we'll likely be facing, given what we do know so far. So ecology is a way to understand the past, to make sense of the past. It's a way to understand the present, and, with some caution, a way of investigating the potential outcomes of the future.

Ecology is not about making people feel overwhelmed or depressed. I believe it's deeply about a sense of wonder. For example, there is the wonder of evolved traits, for instance, of gender. Why does gender exist? Why do things like cooperation, and friendship, and love exist? I like love as much as anybody, but why does affection exist? What are the ecological conditions, and do we see this across other organisms?

Ecology can show us where life is fragile, but ecology is also about life's incredible resiliency. The eruption of Mt. Saint Helens destroyed all surrounding life for miles, and yet within just two years, life was again flourishing even in some of the most ravaged areas.

Now, many of you will remember Prince William Sound in Alaska, the site of one of our country's most serious oil spills. That area has shown great signs of recovery, and we'll report on this later in the series.

You know, I just have to add that it's really quite extraordinary because one of the most really cool, unexpected parts of that was that you have this tremendous amount of oil spilled. Some of it was mechanically picked up; some of it was kind of corralled in booms; some of it was kind of these giant vacuum cleaners that were taking it up, but about half of it was actually taken up by microorganisms that were able to use the oil at metabolic fuel. Absolutely extraordinary.

Now, I also want to confront upfront one fear that people sometimes seem to have about ecology: "What about my self-interest, or my family?" to which I say there is no reason at all to give up your selfish best interest. Remember, life depends on the fact that every living organism will invest in its own survival and perpetuation of its genes. That's selfish behavior, in a biological sense, with a little "s," not a big "S," the little "s." We have to behave and make decisions in our own best interest, and ecology can allow us to see how that self-interest can be expanded into larger scales of recognition than we currently have.

What we can gain from ecology is not a path to some impossible form of altruism. Rather, as informed people we need to ask, and ecology can help us answer, "What is in our enlightened self-interest? What is our deep selfish best interest?" Remember, as a species, if we go extinct, we have extinguished an incredible flame that cannot easily be relit, and most of the species that have ever existed on the plant, 99% of them, are now extinct. We, as a species, have the capacity to understand the complex mechanistic, and sometimes random, forces that shape those outcomes. And so we actually can alter the trajectory and the risk of our own potential extinction.

Now, one of the things that I think is so powerful about ecology, and something I share with my students all the time, is that not everything in ecology and not everything related to positive ecological change has to be really complicated. Ecology can even help us identify what I call the "low-hanging fruit," or at least what are the most important factors for these wide-ranging issues of environmental and ecological decline that we face.

For instance, greenhouse emissions, in our nation, from cattle are almost as significant as those that come from automobiles. You know, what cows do is, they are fermenters, and so they take in large amounts of essentially indigestible grass. And for the sort of situation which we have domestic cattle, they turn essentially indigestible grass into tissues that we can consume. But they do that by having multiple stomachs, and actually, they don't digest their food. They have communities of mutualistic bacteria that live in their gut that actually digest the grass for the cow, but at a price. They produce a lot of natural gas, a lot of methane. And so cows actually produce a significant amount of greenhouse gases. Now, as opposed to what you might think, the gases actually come out of the front end of the cow, the greenhouse gases. They're belching it up.

So there are these interconnections that you might not expect. One of the significant ways in which we can modify greenhouse gas emissions worldwide and in the United States is to address, with open eyes, the amount of domestic cattle that we maintain for food production because that is a significant source. And these are the kinds of relationships that you might not expect when you first think about the kinds of problems that we're facing.

But then, what about the practical implications, or the ramifications, of an enhanced ecological world view? Shouldn't we approach predictions from any science with skepticism? Yes, skepticism is part of sound science, and actually part of our healthy self-interest. In fact, any realm of scholarship requires a healthy level of skepticism. We live in a world right now where we are deluged with data, most of it digital, that washes over us like a rainstorm. And because of the ease and demographication, if you will, of access to the internet, there's a lot of junky information out there that comes at you in very pretty pictures. So it becomes even more of a challenge for our generation, and the generations of young people to come after us, to be even more efficient with their filters as they understand, or attempt to understand, what data are believable and reliable, and which data are so tainted by either lack of competence, or adjusted so much by a particularly extreme world view, that they really can't be trusted. Those skills are some of the most important things that we can impart.

For example, ever since the release of some early greenhouse gas data in the 1970s, which suggested that the Earth was actually headed for a period of cooling, premature, and sort of radical, and wild, public discussion has had consequences ever since. Scientists studying the Earth have arguably become even more cautious than necessary as more data emerged.

The quantity of data done since the 1970s about climate change is so enormous as to be absolutely unavoidable, and we will investigate that and the incredible work of the International Panel on Climate Change, their most recent report, which will be a wakeup call for us in this course. I don't think any ecologist can look at that report and not say that this new understanding that it creates for us has profound implications. Not only for our work as scholars, but for the reality of the stewardship role we play as citizens in the communities in which we live.

It happens to be a very useful coincidence that the cost and availability of oil is increasingly headed for shortages and higher prices at the same time that the issue of climate warming has reached public awareness. It's a bit of a coincidence, but it sure is useful. The conservation and lifestyle changes that people are making in response to petroleum prices actually have positive ecological value to the Earth's sustainability.

Now, the word "sustainability" gets overused quite a bit. It sounds vague, or even empty, but solutions we will need for a sustainable human ecosystem are not without costs.

Now, I'm fortunate to be a co-founder of the Urban Ecology Institute, which is active at the intersection of science and policy. Our mission is to improve urban ecosystems through research, education, and advocacy, and I see this all the time. For example, as part of a coalition, we might recommend to the city of Boston that based on our analysis of tree cover and the structure of neighborhoods that Boston needs to plant 100,000 trees. Well, these trees are not inexpensive. It takes an effort to do this. And even though the trees will hopefully provide cleaner air, and modulate climate, and have tremendous and hopefully measurable social advantages to creating green space, that may mean hiring fewer municipal workers, like some teachers, or fewer police. So if ecologists are making these kinds of suggestions, we better be sure that the data we're standing on are robust.

But, if the physical and social benefits of trees help to make Boston more livable and sustainable, then the investment is sensible. It's in our selfish best interest, especially if the ecological transformation employs people in green jobs and the metrics of public health improve. By the way, "green jobs" may be a term you haven't heard before, but boy, you're going to. One of the areas of tremendous expansion and opportunities is a collection of technical and service jobs across a wide spectrum that are linked to sustainability and environmental improvement.

Now, ecology, as a science, is like that: looking for multiple, mutually supportive, long-term benefits. In fact, as ecologists, one informal way we can tell ourselves that humans have got something right, actually, is when people love where they live. You know, the attitudes people have reflected a very, very wide range of variables. Social demographers and social ecologists have learned that one of the best ways to understand the nature of neighborhoods—we can take a lot of technical measurements that are helpful—is to ask the people who live there what they think about the neighborhood they live in.

And the "where" in which we all live, ultimately, is that very thin, but complicated, realm of the planet studied by ecology. What an ecological diagnosis seeks to offer is, therefore, quite different from the very singular focus on a single cause, and a single outcome, located at a single point in time typically demanded in more traditional science such as physics or chemistry.

After all, what is going on in a particular place at a particular point in time sometimes cannot be understood without reference to a lot that was going on earlier, even a hundred years earlier, or before that, both in time and space. Ecology isn't just applied biology. It is also a different and complimentary way of doing life science. It is holistic instead of monocausal.

Both in this course, and in ecology more generally, we will sometimes find ourselves returning to the same places or to the same examples, but with new questions and with new additional tools of investigation.

Now, ecology has multiple subfields, each of which plays an important role in addressing the interrelated challenges that, taken together, suggest that the living Earth faces a crisis. So there's population ecology, where we look at the dynamics within a single species. Then there's community ecology, where we're interested in the interactions of groups of species and the kind of complexity that emerges from those interactions.

Some of those interactions are behavioral, and that realm of ecology is called "behavioral ecology," and it tends to look at the interactions among animals and the social ecology of that among people. And the reason that we think of behavior as being such a critical metric for understanding the nature of ecology is that, remember, behavior is the most complicated of all phenotypes. It involves the interplay between genes and the environment, all the other scales of biology that go on in living organisms from the cellular, to the tissue, to the organ level. They all give rise to this complex phenotype we call "behavior."

Broadly defined, there's also systems ecology, which includes human impacts in nature, and focus on the really broad interactions that happen at larger spatial scales. Emerging more recently is a new branch of ecology called "urban ecology." Here, we are looking at the interface of nature and human-dominated landscapes, and what are the scales and patterns emerging from this new kind of investigation?

We will use each of these main fields of investigation to help understand the implications of the changes that we observe happening to the Earth's biosphere. Ecology does not operate alone. In this course we recognize and use the contributions of allied fields, such as public health, earth science, economics, engineering, history, and the humanities to knit together a coherent understanding of the intricate dynamics of the Earth.

Now, there are two sorts of changes that suggest the Earth's ecosystem as a whole seems to be in decline—and perhaps you'll need to make up your own mind about this—even in crisis. These changes are in terms of climate and in terms of habitat. Of the two, climate changes are probably more familiar, like sea level rise. In the past 10 years scientists have recorded an average of about 1.8 millimeters rise per year since 1961, and since 1983 that rate has doubled its rate of acceleration. These are all data according to the IPCC report. The sea may displace as many 200 million people in the next 50 years if sea levels continue to rise.

There is a clear and definite increase in greenhouse gases, as has been documented by a whole variety of instrumented and long-term history data, some of the most famous being the Mauna Loa, Hawaii, data by Keeling and his group.

We seem to be seeing harsher and more severe weather patterns related to hurricanes and droughts that may be, and probably are, related to the increased heat energy that's available on the Earth as a function of global warming. Glaciers and ice caps are melting at an alarming rate. The Northwest Passage in the Arctic Circle remained passable for the first time in recorded history throughout the year in 2007.

Habitat changes are perhaps less familiar than the climate challenge. However, habitat destruction is the largest threat to our world's biodiversity, which is why habitat is a central focus throughout this course. Habitat destruction encompasses a range of issues such as deforestation, habitat fragmentation, and even urban sprawl that, together, are a huge threat to our ecosystems.

Now, here are some of the details of the course and what we hope to accomplish. Each lecture is going to cut across physical and temporal scales, like we talk about mercury toxicity. We'll be talking about it from its molecular impacts all the way up to how it moves through the food chain, and why it's so widely spatially distributed, and what are the human social factors that give rise both to its bioavailability within ecosystems and also the cost to humans of that outcome.

The lectures themselves are often coupled in doublets or pairs, with basic ecological principles discussed in the first lecture of each pair, and the human role, or implications, addressed head-on in the second. Lecture Two, for example, continues the introductory attention of Lecture One to the kinds of changes the biosphere and humans are witnessing, but with more direct attention to humans acting as the causative agents.

In Lecture Three, we're going to talk about some of the early history of ecology. Examples include Eugene Odum's great contributions of systems thinking that laid the groundwork for the holistic discipline of ecology to flourish. Lecture Four is a more recent history of ecology, including an introduction to the full range of forces and drivers studied in contemporary ecology.

Now, the vast majority of the lectures are divided into two broad sections. In Lectures Five through Twenty, we will investigate specific forces and drivers that cause ecosystems to change over time, including the flows of energy, materials, including waste materials, water, food, and toxic substances. For example, the large ocean currents serve as great conveyors of energy around the globe and can influence entire ecosystems, such as the ratio of sardines and anchovies in the rich fisheries off the coast of Chile.

The other broad section of the course focuses on ecological interactions and processes, Lectures Twenty-one through Thirty-two. These include microevolution (changes within a given species), population growth, migration, disease, and co-evolution, in which complex systems such as zoonotic disease, where an infectious bacterium sometimes requires a minimum of three other animals just to complete its life cycle. So, for example, over the past 300 years, human land-use practices have reduced the forest cover across the United States significantly, and have also altered the type of forest that is currently observed. Both of these outcomes have considerable ecological impact.

We will conclude with four lectures on human responses to environmental challenges, Lectures Thirty-three through Thirty-six, including current and prospective contributions from science, from new designs and technologies, hopeful directions for cities, and new ways of thinking and living that maintain and restore healthy ecosystems. For example, wildlife parks and reserve systems have been established worldwide. Parks even in developing nations became possible thanks to a complicated economic solution, involving debt relief in return for ecological investment.

Throughout this course, we will do our best to engage you in the world of ecology and enhance the stewardship role that each of us can play in our daily choices we make in our own lives.

In our next lecture, we'll look more directly at the human factor in the overall state of the Earth, and we'll consider, in particular, outcomes that have led to an influential view that humans are driven by self-interest, to enact what has been called the "tragedy of the commons."

So until then, welcome.

Lecture Two
Humanity and the Tragedy of the Commons

Scope:

Although we are one of Earth's newest species, humans have an unprecedented ability to reproduce and modify our environment. We have generally benefited in the short term—with longer life spans, lower infant mortality, and reduced violence—but the legacy costs of technology—such as pollution, habitat destruction, and human displacement—are an ecological burden that has yet to be shouldered.

Outline

I. In this lecture, we look at ecological conditions formed by anthropogenic, or human, influence.
   A. Modern technology is only a couple hundred years old. In that period of time, we have made extraordinary changes in the nature of the Earth and the number of people who live there.
   B. People have benefited in the short term, with longer life spans, lower infant mortality, and reduced violence.
   C. But the legacy costs to this technology are high. We have pollution, habitat destruction, and human displacement.

II. One example of how humans place too much demand on the ecosystem is overfishing.
   A. The increase in demand for fish has caused fishermen to develop techniques to meet those demands. But many fish populations cannot withstand that kind of stress.
   B. Atlantic cod in many ways shaped not only the ecology but also the economy of New England for almost 300 years.
   C. In the 19th century, the Atlantic cod that were caught weighed an average of 100 pounds—and it was not unusual to bring in fish that weighed 200 pounds.
   D. Atlantic cod were pushed almost to the point of extinction, but regulations have helped save them. The numbers being caught are coming back, but their average weights are now down to around 20 pounds because of the impact of fishing.

III. We look at a theory called the tragedy of the commons.
  A. There is an inherent struggle between self-interest and public benefit.
  B. One of the first systems thinkers in this domain was Garrett Hardin, who wrote a piece called "The Tragedy of the Commons."
  C. Hardin's first law states that you cannot do only one thing, because ecological systems are deeply interconnected.
  D. Ecology is about legacies. We need to understand what happened before to understand things as they are now and to think coherently about how things will be in the future.
  E. Hardin stated that in an overpopulated or unstable ecological system, there can be no technical solution, because those systems have high legacy costs.

IV. First we look at a theoretical example.
  A. Hardin used an example of a pasture that is open to as many cattle as possible.
  B. This works well for centuries, but eventually the pasture becomes maxed out and starts to degrade.
  C. In terms of collective wisdom, we know that we should reduce the number of cattle that are grazing in the system.
  D. But each herdsman, even understanding the problem, is compelled to add another cow.
  E. This is because the profit for each cow goes to the individual herdsmen, but the cost of the degraded pasture is shared by everybody.

V. So the question is, how do you solve the tragedy of the commons?
  A. We cannot ask individuals to give up some of their profit.
  B. We need to create legislative intervention, which Hardin calls a form of mutually agreed-upon coercion.
  C. Taxation is an example. We do not enjoy it, but we understand that the individual benefits associated with a group commitment to taxation are worth the investment.

VI. Now we look at a real-world example: habitat destruction in a terrestrial environment.
   A. Habitat destruction is one of the biggest threats to the world's biodiversity.
   B. Through appropriate interventions by governments of developing nations and with incentives from developed nations, habitat destruction can be halted and even reversed.

VII. Increasing human population size and extended life spans mean more resources expended each day and a larger carbon footprint. Back around 1990, we exceeded the capacity of Earth to support the demands that we place on it.

Suggested Reading:

Knox and McCarthy, *Urbanization*.

Levin, *Fragile Dominion*, chaps. 1–3.

McKibben, *Deep Economy*.

Questions to Consider:

1. How does the potential conflict between personal interest and public good impact the current environmental crisis on planet Earth?

2. What is the biggest human-induced threat to the planet?

Lecture Two—Transcript
Humanity and the Tragedy of the Commons

Welcome back. In our first lecture, we had the opportunity to investigate the state of the Earth, and in this lecture, we're going to look at the ecological conditions from an anthropogenic, or human, influence. In this lecture, we're going to talk about overfishing. We're going to look at the theory that helps us to understand why it is that humans appear to act in such selfish ways. We'll look at rain forest destruction as an example of how this theory plays out. We'll think about how much of the impact is linked to population expansion and longevity. Finally, we're going to talk about some of the legacy costs associated with ecological decline. Why it's so high and why it seems to be put off.

One of the most fascinating aspects, to me at least, about the way in which we impact the planet as a species is that we're such a recent species. Modern technology is only a couple hundred years old. In that period of time we've been able to make extraordinary changes in the nature of the Earth and the number of people that live there. We have extraordinary capacity to reproduce and to modify our environment. Our powers of technology and information flow have transformed the post-industrial world.

We're interconnected as a community of humans in so many different ways. People have benefited in the short term, such as longer life spans, lower infant mortality, and reduced violence, but the legacy costs to this technology are high. We have pollution, habitat destruction, and human displacement, and this is an ecological burden, and it really has yet to be shouldered, much less paid off. So in this particular lecture, we're going to focus on the human components of ecosystem change and resiliency.

I want to begin by talking a little bit about this turtle shell that we have here. This is a female diamondback terrapin, or at least the shell of a diamondback terrapin. They're a relatively small, marsh-dwelling turtle that live to be anywhere from 70–100 years old. This is a shell from a female. We've been studying them at our field station on Cape Cod now for about 30 years. We capture them in nets, mark them, we release them, and we use radio transmitters to follow their natural history patterns. This is all stuff we'll talk about later.

But I think for now the important and illustrative point is that this species has been on the Earth for a very, very long time. Turtles precede the dinosaurs. They survived the end of the age of dinosaurs, and yet they find themselves now confronted with a world with cars, and roads, and engine-driven boats, and their populations are starting to decline. This is a federally threatened species, and our work and the work of others is helping to understand how their natural history interacts with humans and what it is we can do to help preserve them as a population.

One example to consider when we think about this whole issue of human impact is to consider the aspect of overfishing. It's an excellent example of how humans place too much demand on the ecosystem. Look, we need food in order to survive. In fact, nutritionists tell us that fish is a great source of protein, so why shouldn't we be getting as much fish as we possibly can?

Well, the increase in demand for fish has caused fishermen to develop techniques to meet those demands. But many of the fish populations can't withstand that kind of stress. In fact, if we thought about the way that we fish, and did so in a terrestrial setting, where we drag nets across the bottom, and we tend to take the top-order predator fish out of the sea, we would be shocked at the way we actually go about the process of fishing. We're going to look at that a little bit later when we talk about trophic dynamics.

But here, we're going to talk about the story of the Atlantic cod. Near and dear to the residents of Massachusetts, Atlantic cod in many ways shaped not only the ecology, but the economy of New England for almost 300 years.

The fisheries in which cod were caught on Stelwagon Bank off the coast of Massachusetts at one time were the richest fishing grounds in the world. But by the late 1990s, the stocks had fallen almost by 95%. Earlier in the 19th century there were nearly 40,000 metric tons of codfish that were harvested from Georges Bank each year. The fish that were captured in that period of time, that were caught, weighed an average of 100 pounds, and it was not unusual to bring in fish that weighed 200 pounds each. You would see them on the docks, and there are images that were captured from the Maritimes that show these big fish being brought ashore, and they were larger than children, in fact. They were as big as adult humans. They were gigantic fish.

Pushed almost to the point of extinction, regulations have helped save the Atlantic cod. The numbers being caught are coming back, but their average weights are now down to around 20 pounds because the population has been so impacted by the activity of human fishing.

Another example, and one that was supposed to essentially solve the problem of overfishing the North Atlantic, is the Chilean sea bass, and yet a similar problem emerges. The species was supposed to save the industry, but population declines due mainly to the way in which they're caught, which are using techniques called "long lines," coupled with illegal fishing, have really decimated the population. These long line techniques that they use can catch seven tons of fish on a single line.

Early on in the explosion of fishing activity off the coast of Chile, reports in the early 1990s showed as many as 34,000 tons of fish caught. By 1995 the capture was down to approximately 9000 tons. So the system had really begun to fail.

A team of researchers led by a fellow named Jackson published recently in *Science*, I think a really important piece of work with respect to understanding our impact on the fishing industry. If I can direct your attention to the figure: Part A investigates the impact of cod fishing; Part B looks at the resiliency of the coral communities; and Part C looks at the impact of oyster fisheries. Those are three very important measures of ecosystem health in the near shore for us, obviously in different parts of the world.

In Part A we see that over a period of time that dates back thousands of years, you can see that we have legacy data on the cod population, and you can see that the mean length of cod that was being caught has dropped dramatically over time as a result of overfishing. We've talked about that.

Part B looks at coral communities over time. It looks at the percent of those coral communities that are maintaining a dominance within the ecosystem. And you can see that those levels are falling as well.

Now, we have two sets of data that are interactive. The first, the solid line, depicts the amount of oysters that are captured off the near shore areas of the American East Coast, and you can see that beginning about 150 years ago there was an enormous increase in the harvesting of oysters. Oysters are considered an incredibly valuable resource for both food and for the marine trade. You can see that there was a peak in the capture that happened about 100 years ago, and it has essentially been declining ever since.

But here's the part, I think, that's most interesting. If you take a look at the dotted line—which admittedly is a little bit complicated because it shows a ratio of planktonic, dibenthic organisms—those are the ones floating in the water versus those that are on the marine floor. That ratio tells us something about the health of the ecosystem, and what it's telling us is that as that ratio becomes more positive, the system is becoming eutrophied, which means it's losing its oxygen level, and it's actually on the way to becoming a dead zone, something we're going to talk about later in our lectures.

Notice that as the impact of humans within this fishery is increasing, measured by the number of oysters that are being landed, you can see that the degree of eutrophication is beginning to climb. Even in the early periods when humans were dredging these near shore environments and mucking them up quite a bit, the oysters were able to keep the ratio at somewhere around 3:1. But as the oyster populations declined, then the europhication exploded, and what we have now is a badly degraded ecosystem that is not easily repaired.

Well, those are data, and I think data are important. Data drive us as ecologists to understand our ecosystem, but ultimately, the importance to us as scientists is that data help us forge a theory. So one of the theories we're going to look at today is called the tragedy of the commons, and it's as much about us as a social species as it is about tools of science.

Now, to understand this environmental conundrum, we have to understand that there is an inherent struggle between self-interest and public benefit. One of the first, early systems thinkers in this domain was a fellow by the name of Garrett Hardin, who wrote a piece called "The Tragedy of the Commons." He was a professor of Human Ecology at the University of California in Santa Barbara. He was one of the first of what we call "systems thinkers," and system thinking is going to play a very important role as we talk more and more about ecology in this series. He was controversial, which is interesting because he was an annoyance to both the political left and political right as a function of some of his views. But his contribution was, nonetheless, quite important.

Let's talk first about what we call "Hardin's First Law," and the First Law states that you cannot do only one thing. If there's a golden rule in ecology, I think that would be it because if we take for our understanding that ecological systems are deeply interconnected—and in some way this runs against the notion of science as we tend to think of it, being very reductionist, isolating our variables, and trying to measure only one thing—in ecology we can't do it because as we have learned and will learn, ecology is about legacies. We need to understand what happened before in order to understand things as they are now, and to think coherently about how things will be in the future.

So Hardin published this article in 1968, very early in our struggles around the issue of sustainability, before the first Earth Day, as an example. He stated that in an overpopulated or unstable ecological system there can be no technical solution. Remember, as a species, we love technical solutions, right? If we use too much water out of an ecosystem, we'll just build a desalinization plant and turn ocean water into drinking water. If we're producing too much nitrogen from waste that we're producing, we'll build a water treatment plant. As a species we're very good at building technological solutions. Technological solutions, however, have high legacy costs, as we'll talk about.

But here's a notion of the tragedy, and it's not the tragedy in the sense of something being terribly sad. It's a tragedy as defined by the philosopher, Whitehead, who said that, "The essence of a dramatic tragedy is not unhappiness. Instead, it resides in the solemnity of the remorseless working of things." In other words, the tragedy is the mechanistic and machine capacity.

So let's use an example to talk about this. First we'll use a theoretical example, and then we'll actually look at some data. As the article was written by Hardin, he used an example of a pasture, a common pasture land, terrestrial pasture, that was open to as many cattle as possible. Now, each individual herdsman has a certain amount of cattle and allows those cattle to graze. This works for centuries, as the resource—in this case the pasture—has not reached its maximum capacity. However, eventually the pasture becomes maxed out, and is now actually starting to degrade.

As a collective wisdom it should make sense, now, that we should reduce the number of cattle that are grazing in the system.

However, this is where Whitehead's comment about the remorseless mechanistic aspect of this comes into play. Each herdsman, even under this recognition, is compelled to add another cow. Why? Well, because the profit for each cow all goes to the individual herdsmen, but the cost of the degraded pasture is shared by everybody. So the negative cost is borne out over many. The profit is taken in by one individual.

This is the same for fishermen that we just talked about. Each is locked into a cycle of boat payments, crew salaries, insurance, and fuel, but there is an escalating value for the fish that are being captured even as the fish are declining in number. As they become more rare, each individual fish becomes more valuable.

It's likely that each act of fishing will be individually profitable even to the end. So, off the coast of my native Cape Cod, people are fishing for tuna using sonar and airplanes in order to capture them because they're so rare, but their value is so high.

So the question is, given the tragedy of the commons, given the reality that exists between selfish best interest and the community interest as a whole, how do you solve this problem? Hardin is a realist in the sense, as he would have described himself, in that we cannot ask individuals to give up some of their profit. What we need to do is create legislative intervention, which he calls a form of mutually agreed upon coercion. What we mean by coercion is that the group decides that this kind of intervention is appropriate.

Now, let's talk about an example of coercion that will be very easy to understand, and that example is bank robbing. Most of us agree that bank robbing is a very inappropriate thing to do, and therefore, there is a very high penalty for robbing a bank. If you get caught, you're in very big danger at the moment you're caught because you're going to be surrounded by armed people, and you're likely to go to prison for having done that. But that's not very controversial because most of us would feel that robbing a bank is inappropriate.

Let's look at something else. Let's look at taxation. Taxation is an example of a mutually agreed upon coercion. We don't necessarily agree to enjoy it, but we understand that the individual benefits associated with a group commitment to taxation are worth the investment.

Let's revisit this notion of how these ecosystems degrade under these ideas. Jackson pointed out that there are different possible scenarios with his investigations of coral, cod, and also oysters. But if we think about this legacy of human expansion, which has led to increased fishing, and then pollution, and mechanical habitat destruction from the types of fishing that we use, like the long lines and the dragging lines, the introduction of species that happen, the impact of things like climate change, what we have are altered ecosystems that look very different than they did then compared to now. So we are stuck with the reality of this kind of change, and the question is, can we use insight from Garrett Hardin's work to alter the kinds of choices that we make?

Let's continue our conversation about the fishing industry. There's a model for the fishing industry where we can compare cod versus lobster. Now, Dietz and his team did some very interesting work. They looked at the cod fishing industry and compared it to the industry that's occurring in the very same geographic region, but admittedly, near the shore, and that's the lobster industry. So the measurements were happening in the Gulf of Maine, and they were measuring cod and lobster.

If you take a look at the chart, you can see from 1980 up to the current time. The solid blue line shows the amount of landings of cod, and you can see that number going down over time. There's a little bit of a peak in the late 1980s and early '90s, but then there's a decline.

In comparison, with the dotted line, through the same trajectory of time, you can see the impact on the lobster industry. You can see that except for occasional declines, the lobster industry has continued to grow. That represents a different historical legacy of regulation. The lobster industry has had a successful mutual coercion, if you will, and the cod industry has not. And so here are two ecological realities that are playing out as a result of different human interventions.

Let's move to another example, and that example is habitat destruction in a terrestrial environment. It's one of the biggest threats to the world's biodiversity. A classic example would be rainforest destruction in Borneo and Costa Rica, both of which have some of the largest amounts of both habitat destruction and critical habitat that needs protecting. For example, Costa Rica, which contains about 5% of the world's known species, is the size of West Virginia. That's an incredible amount of biodiversity, and many species are endemic, which means they're only found in that part of the world.

In 1940, in the middle of the previous century, approximately 85% of Costa Rica was considered to be forested, covered by natural vegetation. By the '60s and '70s that had declined to about 35%. By 1983 it was down to 17%. Clearly, Costa Rica was in a situation of habitat disaster. The land was being cleared to create space for cattle ranching and banana plantations.

Through appropriate interventions by both the Costa Rican government and with incentives from developed nations, today about half of Costa Rica's land is considered forested, and the country is continuing to focus on sustainable practices. We're actually going to look at how some of those practices work later when we talk about debt-for-nature swaps, which are ways in which developed and developing nations can exchange each others' burdens. In this case, the tragedy of the commons was circumvented by regulations that were successful.

On the other hand, we have Borneo. It is the third largest island and one of the most biologically rich areas in the world. From 1984 to 2001 about 56% of the rainforest was destroyed, mostly for hardwood logging and for palm oil. The World Wildlife Fund keeps a Living Planet Index, which helps them to monitor the conditions of ecosystems around the world, and we'll be talking about these living planet indices throughout this series of lectures. If you take a look at the Tropical Forest Index, which measures a whole variety of factors associated with their ecological resilience, you can see from 1970 to the present the Living Planet Index for tropical forests has gone down dramatically. It's one of the ecosystems that we have the most challenge in maintaining. I think it's interesting because palm oil was a solution to a problem that only emerged recently because of an insatiable market for fast food and prepared foods that can be produced with palm oil, and even more recently as a tool for making biofuels.

Palm oil production claims a large amount of land and is detrimental to the surrounding ecosystem because it is a monoculture. As a monoculture it creates a very, very inhospitable environment. To top that off, fire is used to clear the land. It creates air pollution and significant problems for the surrounding areas. In fact, if you look at the impact of burning the lands, destroying the soil, producing the palm oil, transporting it, actually, the carbon savings, if you will, by using that as a biofuel is a net loss with respect to ecological resiliency. But that's a conversation for later.

Much of our human impact is linked to population growth and modern technology. That's a critical take-home point that we'll be returning to over and over. In developing nations, this notion of modern technology is enabling humans to live longer and have a higher survivorship at birth, which is good. If we think back to medieval Europe, living to the age of 30, which would be a high average, depending on gender and social status, was considered to be successful. Even early in the 20th century, living to be 40 was considered the average. The current world average is now 67, which ranges from a high of 82 in Japan to a low of 41 in Sierra Leone, a war-torn and poverty-stricken human ecosystem.

There is even significant variation in the United States. We do enjoy a relatively high standard of living and long lifespan. The longest average lifespan is in Hawaii, where the average is 80. The District of Columbia is the lowest in the United States, which is 72.

An increasing population size, however, and people living longer, means more resources expended each day and a larger carbon footprint, something we're going to investigate and measure later on. An increasing ecological footprint creates additional stress on the environment. Returning to this Living Planet Index for a moment, we can take a look at this chart of world biocapacity as was measured from 1960 to the present, and we can see that if we look at biocapacity as a measure of how many Earths it takes to support the activities that we have—and again, this is a very clever analysis of multiple variables that help us determine what the carrying capacity of Earth is—you can see that somewhere in the vicinity of around 1990 we exceeded the capacity of our Earth to support the demands that we've placed on it for the long term. This is one of the examples that I brought up when I talked about this burden that we are shouldering as far as ecological impact, that we haven't been able to repay that debt yet.

Now, human populations have benefited quite a bit, at least on the short term, with respect to technology. However, the legacy costs are high. We've drained and filled wetlands, and we've seen that as a quick fix for transportation issues with respect to crowding of coastal cities. Boston has lost 95% of its historic salt marsh. Only about

350 acres remain. As a result, the coastline is battered by relatively minor storms.

Development and dredging have claimed nearly 2000 square miles of marshes along the Louisiana coast for homes and petroleum transport. However, we now know that salt marshes and barrier beaches are critical buffers to storm surge. Dauphin Island, which is a barrier beach off the coast of Alabama, was slammed first by Hurricane Lili in 2002, then Ivan in 2004, and both Dennis and Katrina in 2005. Now, the barrier island did exactly as it was supposed to. It absorbed the storm surge and reduced the impact to the mainland, but houses and property on the island were decimated. As a result, the barrier islands and beaches were still there. The homes were destroyed, and the houses should not have been permitted.

Again, this is a classic tragedy-of-the-commons example. Individual homes may have a beautiful view and high property value, and be of high value to the individuals. Their losses are covered by insurance companies, and the environmental costs are covered by the government. So, you have a system that is inherently out of balance.

The legacy costs for ecological decline are very high. Now, Wilson and a team have done some work where they have investigated both the measure of ecological decline and compared that to the expansion of activity. In other words, human-based activity on the planet, and they see that the two forces are essentially in opposition. As we have increased our gross product of activity and technology—that's a line that keeps going up—the impact on the planet, the planet's resiliency, has essentially decreased. Sometimes these impacts are unintended, but they're just as important.

I'll close with a story about European rabbits that were introduced into Australia in the middle of the 19th century. Australia, as a continent, has mammals, but they're a more primitive form of mammals; they're marsupials. When the Europeans, especially the British, colonized Australia, they wanted to bring rabbits over because they had a fondness for rabbit meat and for rabbit hunting. Only a few rabbits were brought at first, a couple of dozen. They were brought in 1859. Some of them escaped, and the extraordinary thing is, by 1928 there were 500 million rabbits loose on the continent of Australia. They caused massive land erosion and vegetative destruction that lasts to this day.

Now, to control the population they introduced the Myxomia virus that had been found in the European rabbits. It had a devastating effect on the Australian rabbits. It killed 90% of them. But within 20 years after that the population was back to 300 million, and now these are sort of super-rabbits that are able to withstand these kinds of viral interactions. It's had the effect of pushing the Australian rabbits, the greater and lesser bilbies, to the point of extinction.

We've seen similar patterns around the world. In our own country, plants like purple loosestrife and Phragmites have devastated fresh and saltwater marsh systems. They were originally brought here as ornamental plants. Zebra mussels got to North America in the bilges of boats and are now devastating the Great Lakes.

I don't think any of us can live in a house anywhere in North America without a house sparrow as a neighbor. House sparrows are actually finches from Europe that immigrated to the United States along with western settlers who came over at the same time. So this is an example of how ecosystems change as a result of human social forces.

In the next lecture, we're going to continue this look at Hardin's Legacy, but we're going to consider the human impact this time from the standpoint of the research and ecology that have brought us the modern science. So until then, thank you.

Lecture Three
Ecology—Natural History to Holistic Science

Scope:

In this lecture, we look at the many levels on which we explore ecology as well as the way ecology intersects with other contexts. We also discuss several key contributors to the field, stretching back as far as Charles Darwin, before looking at how modern technologies are allowing ecologists to investigate subjects in new ways and on new levels.

Outline

I. It is hard to unify a science as wide ranging as ecology.
   A. Ecology's first big unifying hurdle was to distinguish itself from natural history.
   B. Natural history is essentially a contemplative process, and ecology as a science is a tool for prediction.
   C. Charles Darwin turned the art of observing nature into a powerful scientific tool and a holistic explanation of how biological systems work.

II. Ecology has long been a bit of a struggle between various conflicting ideas.
   A. A holistic presentation of the ideas must have an ecological context, but also socioeconomic and institutional contexts.
   B. Where these contexts come together, we can have the strongest partnership in the development of ecological ideas and in using ecology as a tool for human betterment.

III. Let's explore the levels on which ecologists do their work, from the most fundamental to the most complex.
   A. On the molecular level, ecologists investigate interactions among the chemicals associated with living systems.
   B. On the cellular level, we begin to investigate the first scale of autonomous living entities. The field of microbial ecology focuses on this level.
   C. On the level of tissues, we focus on the uptake of toxic chemicals, which can have significant downstream impacts.

- **D.** On the organ level, we look at the evolution and adaptation of organs to different ecosystems.
- **E.** Organisms are the most important unit of evolution because it is individual organisms that carry DNA. On this level, we look at things like the impact of human behavior on other organisms.
- **F.** The next level is that of populations; this refers to interactions among individuals within a species.
- **G.** The community level is where we study interactions among multiple species. This is one of the most important domains of ecology and is the core scale for understanding interactions.
- **H.** Ecosystems, which include communities and the physical environment, provide a large-scale regional view of the ecological health of a landscape.
- **I.** The largest scale is biomes, which are large regions of the world defined by water availability and vegetation.

IV. Key contributors to the science of ecology.
- **A.** Charles Darwin and Alfred Russell Wallace investigated the origin of species, which in many ways was the birth of community ecology.
- **B.** The idea of natural selection actually gets its roots from Thomas Malthus, who developed mathematical models for understanding populations.
- **C.** Ernst Haeckel is credited with coining the word "ecology," which comes from the Greek word *oikos*, meaning "house."
- **D.** Eugenius Warming transformed ecology into a science through his work in botany.
- **E.** Victor Shelford helped to found the Ecological Society of America and is best noted for his law of tolerance.
- **F.** Arthur Tansley was the founder of British Ecological Society. He rejected Lamarckism and was a fully modern Darwinist.
- **G.** Eugene Odum is best remembered for thinking about nature as a set of interactive systems. His book *Fundamentals of Ecology*, first published in 1953, was the first and only textbook in the field of ecology for decades.

V. Modern ecology accounts for ecosystem services, especially in arenas like urban ecology and landscape ecology.
   A. Modern ecology attempts to interweave the ideas of ecology and human survival.
   B. Examples include the long-term ecological research sites that have been set up in collaboration with the National Science Foundation and about 30 research teams across the country.
   C. Patterns of human activity affect the ecosystem in such things as demographic patterns, economic systems, power hierarchies, land use and management, and the designed environment.
   D. The changing world challenges ecologists to expand the nature of our science.

VI. The emergence of technologies that permit investigating animals and plants remotely, delving into their genetic secrets, and reconstructing their natural history through chemical analysis is changing the way that ecologists work.

Suggested Reading:

Leopold, *A Sand County Almanac*.

Sutherland, *From Individual Behavior to Population Ecology*.

Questions to Consider:

1. How do the ideas of legacy and ethics shape the modern ideas of ecology?
2. What are ecosystem services, and how do they influence the science of ecology?

Lecture Three—Transcript
Ecology—Natural History to Holistic Science

Hello, and welcome back. In the last lecture, we were investigating the impacts that humans have had on ecosystems. Today, it's a transition. We're going to be shifting from thinking about humans as forces that change the planet to the science that helps us to understand how we can measure and evaluate those kinds of forces.

Now, ecology has different scales, very much like history. It involves legacy, such as the land ethic. But today we're going to focus on key contributors. We're going to talk about Haeckel, who coined the term "ecology" from the word *oikos*; Eugenius Warming, who brought early ecological thinking to the science of botany; Arthur Tansley, who introduced the whole idea of ecosystems; and, of course, Darwin, who glimpsed the big picture of how systems change over time, focusing on natural selection as an explanation for life's unity and diversity.

I want to begin with some really cool imaging that we have from our laboratory. What you see here in this sort of grainy, gray image is actually a thermal image of the deer that we studied. What you're seeing from the deer is the energy that's given on in the infrared, and this is one of the modern tools that we can use to measure the number of animals in a system. Now, we're studying white tailed deer as a function of trying to understand the Lyme disease system. There's a relationship between deer density and human risk of Lyme disease. We're going to have that conversation in detail a few lectures from now.

But you can see here these images were taken at night, and we wouldn't be able to see into that vegetation to be able to count those animals. But yet because they're mammals and they're giving off heat, you can see that we can count them. You can actually see birds in the background and other organisms as well. I can only imagine the kind of conversation we could have with Charles Darwin if he had had those tools to be able to assess animal density.

In some respects, ecology has been a bit of a wrestling match among conflicting ideas, a struggle to emerge from this shadow of natural history, and a bit of a feud among the zoologists, botanists, and microbial ecologists for a holistic vision that is really the hallmark of this discipline.

Like so many sciences, we have powerful tools to investigate, and theories to help explain, the details of particular phenomena. However, it is much harder to unify such a wide-ranging science than it is to explain its individual parts. Like the very root of the word "ecology," *oikos*, which is Greek for the idea of a house. So now, imagine all the individual systems within a house—the electrical system, plumbing, heating, the house construction, and human design elements. How do these interact to form the whole system? Surely they do, but we consider them all as subsystems. Hence, when a new house is built, we have separate inspections from the plumbing inspector, and the electrical inspection, and so forth. Yet, we live in a house as a whole system, perhaps only envisioned as a complete system by the architect during the design and construction phase.

Now, architecture has the model of something called "universal design," which is a theory of construction that considers the challenge of providing physical access to all parts of a building's systems for all people, regardless of their physical limitations. But this is still a fairly focused and narrow vision.

Ecology's first big unifying hurdle was to distinguish itself from natural history. Robert Peters, in an essay published in *Perspectives in Biology and Medicine*, draws the distinction between natural history and the science of ecology, in that natural history is essentially a contemplative process, and ecology as a science is a tool for prediction. He argues for a retreat from natural history, other than to formulate initial impressions of an ecosystem.

However, not all ecologists accept this premise. E.O. Wilson, one the 20th century's most accomplished ecologists, argues that natural history, the uncovering of new species and their interactions, is at the very core of the ecological science and is absolutely integral to moving the science forward.

As you can see, the debate is healthy and, generally, a science grows stronger as a result. So let's begin our conversation with one of the world's great naturalists, Charles Darwin. Now, really, any conversation of Darwin is really an homage to the movement of naturalist to science. He turned the art of observing nature into a powerful scientific tool and holistic explanation of how biological systems work.

But ecology is a relatively recent science, and has grown out of Darwin's work, and ecology is understood at different scales. If you take a look at this graphic we have provided for you to help lead you through this idea, you can see that when we think about this science, when we think about understanding a holistic presentation, if you will, of the ideas, we have an ecological context for our understanding. That's the data, the mathematical models, and the scientific responsibilities that go with doing this kind of science. But there's also a socioeconomic context with values, and interests, and human information. There's the land and these other assets that are also in essence a responsibility, if you will, to understand how the system operates.

And then, of course, there's an institutional context because we're dealing with humans. Here we have law, and authority, and public sector responsibilities. So these overlap and help to create this holistic understanding. Some of the overlap involves a regulatory overlap, and some of it involves the social obligations that as citizens and stakeholders we bring to a conversation around ecology and its use in ecosystem protection. There are informal decisions that we make all the time around our behavior in the communities in which we live. But where all three of these come together we have the potential for the strongest partnership, both in the development of ecological ideas and ultimately, using ecology as a tool for human betterment.

So when we think about ecology, we think about these levels developing from the most fundamental to the most complex. Let's explore some of these levels in which ecologists do their work.

First of all, ecologists work at the molecular level, where we are investigating interactions among the chemicals associated with living systems. An ecologist here might be investigating things like pesticide residues. We know that these residues have tremendous impact on things like amphibian development, which we're going to be addressing later in this lecture series. Some of these residues actually mimic the endogenous hormones that organisms have, and send these haywire signals in the developmental regime of frogs. As more data are coming out, we're discovering that humans are susceptible to these as well. In rural areas, there is actually the emergence of an understanding that certain months of the year when humans are conceived, especially if it's in the late spring and early summer when pesticide residues tend to be highest within ecosystems, downstream impacts of developmental abnormalities in developing fetuses are actually highest during periods of time when conception happens in that window of the calendar year.

In addition, ecologists work at the cellular level. Here, we begin to investigate the first scale of autonomous living entities, such as the soil bacteria, and the photosynthetic algae. These organisms have tremendous impact on nutrient cycling or greenhouse gas levels. In fact, there's a whole field of ecology called "microbial ecology" that focuses at this level. Some of our most powerful theories and models that are driving our understanding of ecology come from this scale.

Moving up in scale, we have the tissues, and here we're seeing cooperative interactions among cells, primitive living societies, if you will, and you may not have actually thought about the tissues in your body as being a cooperative co-venture among individual cells, but they are. Here, we ecologists focus on the uptake of toxic chemicals, certain compounds of which are sequestered in certain tissues and have significant downstream impacts, like lead in the nervous system, and mercury in the tissues. We're going to be talking about both of those later on in the course.

The largest scale yet, organs, which are structural and highly specialized. Here, the evolution and adaptation of organs to different ecosystems, understanding the ecology of carnivores versus herbivores, which changes dramatically the body structure and behaviors that these animals go through.

And then there's ecology at the level of organisms, which is the most important unit of evolution because, as we'll see in later conversations, it's individual organisms that carry these bags of genetic information, the DNA, and those bags either get transferred to the next generation or they don't, fundamental measure being that of fitness, and we're going to talk about that. Here, things like the impact of urbanization and human land use change on animal behavior, and animal foraging strategies has a huge impact in understanding the nature of ecology. The effect of lights on bird migration and sea turtle orientation is another piece of work done at this scale.

Another scale increasing in size is that of the ecology of populations, and these are interactions among individuals within a species. Here, we have the considerations of things like local and global extinction events; gray wolves, which are locally extirpated but still have a global presence.

The next level is communities. These are multiple species interactions, and this is one of the most important domains of ecology. It's the core scale for understanding interactions in ecology, the movement of energy, the cycling of materials.

Now, ecosystems, which are communities along with the physical environment, provide a large-scale regional view of the ecological health of a landscape, such as measuring forest health in urban areas, as well as coastal fisheries, both areas that we will invest heavily in, in this course.

Even larger, we have the biomes, which are large regions of the world defined by water availability and vegetation. Here, we're considering the entire terrestrial or aquatic and atmospheric living system. And here we're connecting to things like Lovelock's Gaia hypothesis, and we will talk about biomes and biosphere later in the course.

Now, the early ecologists focused on something called "limiting factors." The world of animal ecology and plant ecology were generally quite separate.

We begin our conversation with Charles Darwin, and we need to include Alfred Russell Wallace in this conversation as well. Their contribution was to think about the origin of species, which in many ways is the birth of community ecology. Today's species are derived from ancestral species in an unbroken lineage of genetic transmission. Remember, a broken chain equals extinction. And natural selection is their model of what drives that change, evolutionary change in populations where favorable heritable traits are inherited and become more common over time. Organisms that survive pass on those traits to their offspring. Over time, the variation among members of the population increases and the characteristics change, i.e., they evolve.

So we define natural selection as the force that shapes the genetic characteristics of populations, favoring the most reproductively successful combinations, and therefore, certain individuals.

Now, this idea of natural selection actually gets its roots from an English clergyman and fledgling economist, Thomas Malthus. He lived from 1766 until early into the 19th century–1834. Although he had a career as a clergyman, he studied mathematics at Cambridge, and he grew up in an incredible environment. His family was friends with the philosopher David Hume, and Jean Jacque Rousseau. I can just imagine the kinds of conversations that happened around the dinner table when he was a young boy.

He developed mathematical models for understanding populations, and wrote the essay called *The Principal of Population*. From this we get the idea of the Malthusian catastrophe because his calculations suggested that the Earth's human population would double every 25 years without some kind of external population control. These ideas tremendously influenced Darwin and, in fact, influenced other scientists like Verhulst, who developed his logistic growth models after some of the ideas that Malthus had forwarded.

Unfortunately, Malthus is remembered in social history for his relatively cruel philosophies that emerged from his belief that the Earth was becoming drastically overpopulated. It had a tremendous influence on British social policies, including those that led to the Irish potato famine, which in fact, we will talk about later from an ecological perspective in this lecture series. In fact, Malthus's pessimism gave rise to the idea that economics is the "dismal science." But his contributions were really helpful in forging an idea about the way in which populations grow.

Now, Ernst Haeckel actually is credited with coining the word "ecology," which comes from the Greek word *oikos*, meaning house. He supported the idea that an individual's ontogeny maps its entire history or phylogeny. In other words, an individual's development actually goes through sort of a mini-evolutionary process.

He was a zoologist, and he studied comparative anatomy. He was an aggressive promoter of the ideas of Charles Darwin. He dabbled in the studies of human evolution, making wild predictions without any evidence. In fact, some of his predictions are poked fun at today because not only do they tend to be wrong, but they had some pretty heavy social overlays. He predicted that humans evolved in the East Indies. In the end his theories were weakened by his belief in Lamarck's theory of acquired inheritance, that individual organisms can evolve within their lifetime and then pass those on to their offspring. It was a compelling idea at the time, in which evolution and natural selection were getting their scientific sea legs, eventually proved to be wrong, but there were a number of scientists who made significant contributions, but their vision was a bit crippled by the fact that they couldn't let go of the Lamarckian model.

Another critical contributor to the idea of ecology is Eugenius Warming. He transformed ecology into a science through his work in botany. He believed in the direct descent of living organisms in the same species, and he incorporated some of Darwin's ideas of natural selection. But he remained at heart a Lamarckian, and so his initial contributions in plant biology were very, very helpful, and really was one of the foundations of the area of ecology that botany would occupy. But again, his vision was limited by this Lamarckian model.

Victor Shelford came along. He was an American zoologist and an animal ecologist. He did most of his work at the University of Illinois, and one of the things he's most remembered for is he helped to found the Ecological Society of America, which is a unifying organization to which most of us as ecologists belong. One of the things he did is, he was in charge of the Illinois natural survey from 1914 to 29. By the way, this natural survey became the model for the national, federal, efforts to understand the natural history and ecology of our landscape.

He's best noted for something called his "Law of Tolerance," where he says, "A species' distribution is determined by the environmental factor for which the species has the narrowest of tolerance," such as oxygen, or pH, or salinity for something like marine organisms.

Back on the botany side, we have Arthur Tansley, who was an English botanist, who was championing this idea of ecosystems. He was the founder of British Ecological Society, and these two ecological societies—Ecological Society of America and the British Ecological Society—have been extraordinary forces uniting and bringing together ecologists at annual meetings and regional meetings to help sort of flush out the core aspects of the science. He was heavily influenced by Warming, but he rejected Lamarckism. So here we have a fully modern Darwinist, or Darwinian view of evolution being incorporated into our thinking about ecological systems.

Warming considered that there was an interaction of the biological and the geophysical forces. Later in life he branched into psychology and had some very interesting ecological ideas that helped to push forward that discipline as well. He did important research on species range and looking at environmental conditions. In 1939, he published a huge map of the vegetation on the British Isles.

Now, we begin to understand the systemic approach through his looking at the distribution of plants from an ecosystem level, sort of stepping back. I can only imagine if he had GIS capacity to create these beautiful digital maps what his vision would have done.

Switching back to the United States, we consider the work of Eugene Odum, who lived from 1913 to 2002, 88 years of extraordinary productivity. He's best remembered for thinking about nature as interactive systems. One of his contributions was that he wrestled ecology away from the idea that it was just an additional branch of biology. Instead, he envisioned ecology as an integrated free-standing science that investigated the interactions of living organisms and the abiotic Earth across all scales. He did some great pioneering work at the Savannah River Installation in Aiken, Georgia, and he began, actually, with a government grant to investigate the ecological implications of the nuclear weapons production that was taking place on the river ecosystem. His early career was frustrated by his colleagues, who misunderstood ecology as just a formalized type of natural history, devoid of any unifying themes. Like so many great inventions, if you will, in human history, frustration and need becomes the driver for change. So Odum sought to correct this with his most famous work, his book *Fundamentals of Ecology*, which was first published in 1953 and updated many times later. It was the first and only textbook in the field of ecology for decades.

Odum spent most of his career at the University of Georgia, and he was a steadfastly critical thinker about the science of ecology. He helped to really sharpen the rigor of ecological theory, and he pushed to separate the politics of the environment and the science of ecology apart so that the science could stand as a body of work and, therefore, be useful in the policy and political arenas. But the science would be driven by theoretical questions. We could argue that he was really one of the first scientists to study ecology on this larger scale.

Now, some of the key ideas of ecology involve legacy and land ethic. Here, we see some of the historical development of environmental ethics, like The Golden Rule, the relationship between individuals and their society. The sage, Hillel, formulated The Golden Rule as a precedent for understanding early Jewish moral law. "That which is hateful you do not do to your fellow."

Emerging from this idea was Aldo Leopold's' *Land Ethic*. He considered the relationship between *Homo sapiens* and the living Earth in which we are co-habitants. Some of his most important essays included *Thinking Like a Mountain*, in which he writes of the interaction between the deer that live in that environment and the role that they play in the ecology of the mountain. It was really transformative work.

He also wrote *A Sand County Almanac*, which was actually published in 1949, after his death. He worked to forge a union between ecology and conservation, understanding that ecology was a distinct science, but felt strongly about the use of ecology as a tool for human betterment. So his major contributions worked towards soil conservation in his native Wisconsin, which led to the state's first Soil Conservation Act. He was a co-founder of the Wilderness Society. He worked for the U.S. Forest Service and the University of Wisconsin, and stands as a landmark of the kind of work that can transform a science into a useful tool of human improvement.

Modern ecology accounts for ecosystem services, especially in arenas like urban ecology and landscape ecology. And really, modern ecology is a variation on the theme of "what have you done for me lately?" This links the science of ecology to human survival. Modern ecology attempts to weave these two ideas together.

One example would be the long-term ecological research sites that have been set up in collaboration with the National Science Foundation and about 30 research teams across the country. Interestingly enough, two of which are actually set up in cities: Phoenix and Baltimore. They have a human biological focus, and they're led by a new wave of ecologists, Nancy Grimm in Arizona and Stewart Pickett in Maryland.

If we take a look at the way in which they are considering the use of ecology as a tool, you can see that they've created this integrated social ecological system, where there are social components like the demography, and technology, and economic structure, and institutions that humans bring to the landscape. And there are these bio-ecological components that have been traditionally measured in ecology, like primary production, and biodiversity, and nutrients, and disturbance, and these come together through an emergence of our understanding of the interactions humans have with their environment. How do they use land? What are the land cover implications? What about consumption, and disposal, and so-called "flux," the way in which elements that support an ecosystem change over time? And then there are these external systems that drive that: biophysical, and political, and economic forces that shift these ecosystems. It's a new and integrated way of understanding the way in which ecosystems function. We will get into this in great detail later on in this course, but I want you to understand that one of the great transitions in modern ecology has been to bring in the human factor. These projects consider humans as central to understanding ecology.

What are the demographic patterns? What are the economic systems? What are the power hierarchies? How does land use and management impact the designed environment? This takes us back to Tansley. Tansley was thinking about this complicated relationship between the physical environment in which organisms live, and the biotic environment, those are the organisms themselves. That's Tansley's model. Tansley would recognize, if he were alive today, the contributions that—let's call them the urban ecologists— the urban ecologists are making to our overall view of ecology by simply adding additional boxes. Tansley would be comfortable with this biotic box, and the physical box, and the interactions that take place between the two of them. People like Stewart Pickett and Nancy Grimm would say, you know what, there's a significant part of the environment that has been built upon by humans. It turns out that it's only about 6% of the world's land mass, but it happens to be the most valuable pieces. There's also a social force associated with the way in which humans live. So we know that humans interact with their built environment. If we take those two ideas and attach them to the original Tansley model of the biology box and the physical box, and then we add the built environment and the social drivers and complexity that humans bring, we now actually have captured all of those apparently disparate ideas that give rise to our understanding of ecology. And as we think about the changing world and the increase in the human footprint on the landscape, and the fact that cities as emergent systems are the most important ways in which humans live, it really challenges us as ecologists to expand the nature of our science. Maintain a robust science, but use that science in support of human need.

So looking forward, what does ecology look like as we move into the next generation of scientists? Andrew Read from the University of Edinburgh and James Clark from the Nicholas School for the Environment at Duke in a recent edition of the Journal, *Trends in Ecology and Evolution*, were asked to predict the future of ecology as that journal celebrated its 20th year of publishing.

The authors were quick to point out how dangerous it was to predict the behavior of scientist. However, what they saw was a widening of spatial scales where ecology will contribute, especially this unification of microbial ecology with its theoretical and modeling base with the powerful natural history data of the larger animal and plant ecology.

They also envisioned a blossoming of applied ecology—science in service to human needs—with a focus on climate degradation and habitat loss. They also suggested new tools will permit entirely new kinds of questions to be asked.

The emergence of technologies that permit investigating animals and plants remotely, delve into their genetic secrets, and reconstruct their natural history through chemical analysis is forever changing the way that ecologists do their work. Satellite-based radio telemetry permits scientists to track the movement of animals to a very detailed degree. Polar bears, wolves, whales, even small animals can be fitted with GPS radio collars and followed remotely, allowing scientists to get very detailed information on their geospatial activities. Our lab uses radio telemetry on coyotes and turtles to understand more of their natural history patterns and behavior. We're going to talk about that a little bit later.

Molecular biology has also transformed the science of ecology, allowing researchers to track the lineages and evolutionary patterns of organisms in ways thought impossible just a decade ago. For example, genetic analysis is unraveling the mystery of isolated populations of organisms. Seaside sparrows, which were considered to be at least six subspecies, sort out to only two distinct populations when the appropriate genetic markers are used.

The endangered red wolves, which have shown some hybridization with coyotes, present a population tangle that has implication for management. Researchers at the University of Idaho, along with U.S. Fish and Wildlife Service scientists, have located at least 18 gene loci that are unique to red wolves, and that makes it easier to differentiate populations of wolves and coyotes along their zones of hybridization.

The emergence of chemical isotopes as a way of investigating ecosystem function has also transformed our ability to reveal trends in ecosystems. As you probably know, isotopes are forms of an element that vary in their number of neutrons. They're functionally identical in living systems—these isotopes—especially of carbon, nitrogen, oxygen, and sulfur. We can track those in the tissues of organisms. Different ecosystems have different ratios of isotopes, and so samples of tissue can be traced to the areas in which they were formed, very useful if you are trying to figure out where a bird molts along its migratory route, for instance.

In the next lecture, we're going to focus on the forces and drivers that change ecological systems. This is the beginning of ecology as an integrated process. Thank you.

Lecture Four
Ecology as a System—Presses and Pulses

Scope:

Where ecologists used to focus on stability, we now have a new model for understanding the dynamic nature of change. This model allows us to examine short- and long-term events (called pulses and presses) while focusing on both earth science and human impact.

Outline

I. This lecture shifts our focus back to ecology as the study of systems.
   A. The drivers, or forces, that change ecosystems can be biological, geologic, physical, and social.
   B. We can envision ecology as the result of a variety of legacies, which include both predictable and contingent outcomes.

II. In the 1970s, a new approach to ecology emerged, based on constant states of change and the recognition of humans as the critical ecosystem drivers.
   A. In the human ecosystem, the social complex, the built complex, the biotic complex, and the physical complex interact.
   B. From studies in the early 1960s, ecologists' shifted their focus from stability to resilience, which is the ability of an ecosystem to undergo disturbance and maintain its functions and controls.
   C. We can divide forces of change into those that occur in short doses versus over long periods of time.
   D. The new model for understanding this dynamic nature, the Integrated Science for Society in the Environment (ISSE) framework, includes short- and long-term events and integrates both earth science and human impact.

III. The drivers that impact ecosystems take place on a whole spectrum of levels.
   A. On the global biophysical level, we have things like ocean currents and trade winds.

- B. On the local biophysical level, we have forces like primary production, nutrient cycling, and the microclimate.
- C. Geologic drivers include the topography and shape of the landscape, and how that landscape is changing over time.
- D. On the local geologic level, we have morphology, which is the shape of the land and how water is related to that system.
- E. Human social forces include power structures, human migration, and human events.

IV. There are 5 central questions that help us understand the relationship of humans to ecosystems.
- A. How do long-term press disturbances and short-term pulse disturbances interact to alter ecosystem structure and function?
- B. How does the distribution of organisms we find in an ecosystem either cause or serve as a consequence of some of the other dynamic aspects we are measuring?
- C. How do altered ecosystem dynamics affect ecosystem services?
- D. How do changes in vital ecosystem services feed back to alter human behavior? (Do we learn from our mistakes?)
- E. Which human actions influence the frequency, magnitude, or form of press and pulse disturbance regimes across ecosystems, and what determines these human actions?

V. External drivers that impact ecosystems can be divided into presses (long-term impacts) and pulses (short-term impacts).
- A. Presses include drivers taking place over centuries, such as climate change, forest succession, coastal erosion, and human social movements.
- B. Pulses are often catastrophic events, such as hurricanes or tidal waves, but also include disease and invasive species.

VI. It is important to note that not all disturbances are detrimental; some are actually central to ecological resiliency.

Suggested Reading:

Collins and Wallace, *Fire in North American Tallgrass Prairies*.

Gonick and Outwater, *The Cartoon Guide to the Environment*.

Questions to Consider:
1. From a modern ecological perspective, what are the concepts of dynamic change and resilience?
2. What are the key forces and drivers in ecosystems?

Lecture Four—Transcript
Ecology as a System—Presses and Pulses

Hello, and welcome back. In our previous lecture, we had investigated this notion of legacy, and begun to investigate not only how ecosystems are formed and impacted by history, but also looked at the history of the people who investigated some of the early aspects of ecology. But ecology is the study of systems, and it's the systems thinking part that we need to re-engage today. You know, all ecosystems and the organisms that live there are shaped by the forces of natural selection, and ecosystems are always changing. The drivers, or the forces, that change ecosystems can be biological, geological, physical, and social. In fact, in the next two lectures it's going to take us a while to unpack all of these ideas, but we're going to start today.

I want to begin with an image that is very dear to me. It's a picture of a scrub pine taken from our field station down on Cape Cod. It was taken in the early morning, and you can see that there is water dripping from the pine bough. I think it's a great way for us to begin to think about this notion of history and connection to systems. Cape Cod, for those of you who may have visited, know that it sticks out in the ocean. It has some of the most challenging weather patterns of any place in North America, and the organisms that live there have to be particularly hardy because, for the most part, the soils are relatively poor; they're wind-driven; it's mostly sand dunes. And so the creatures, especially the plants, that live there are pretty hardy. Scrub pines are particularly scraggly-looking plants, and many of the people who live there actually think that they're ugly because sometimes half the tree will essentially have no pine needles on it at all. That's usually the side that's facing toward the wind.

But I look at those trees and I think survivors. I think organisms that are able, really, to withstand extraordinary impact. In this particular image you see this sort of gentle and soft morning that suggests that there's a calmness and quiet to the ecosystem. In fact, before the Sun has risen, the dew in the morning is still hanging off the leaves. The day will get warmer, the wind will pick up, the system will become more energized, but this is sort of a quiet moment. It's very reflective for me to see that. It's quite an extraordinary place.

Remember from our history when we were talking about Arthur Tansley's work in the 1950s. It really helped develop our understanding of examining ecology as a system. You know, we can envision ecology as the result of a variety of legacies, and these legacies include predictable outcomes, but also contingent outcomes.

Now, remember Tansley's idea of these multiple boxes. We had the biotic complex and the physical complex, and they were interacting. So we had nonliving forces and organisms that were living there, and they impacted each other. This was an important idea that Tansley brought forward.

More detailed research that we will be investigating in more detail later in our lectures by people like Stewart Pickett, Nancy Grimm, and others, suggests that there is an important additional component. We touched on it just a bit in the last lecture, and that's thinking about the social aspect of what humans bring to ecosystems, and the fact that humans build things along the way. We used the term "ecosystem engineers" in our last conversation, and that's an important component when we're thinking about this diagram or this conceptual framework for our work in this lecture.

So we have the social complex, we have the built complex, and then we have this box of the biotic complex, and the physical complex that Tansley provided for us. All four of these are interacting, and so while historically it may have been enough to think about the storms and the geology of a system as it impacts on the organisms that live there, we really now need to be asking additional questions. What is the impact of the humans that live there, both in the humans with respect to their behavior, the so-called "social complex," and then what have humans physically done to the ecosystem with respect to building roads, or diverting rivers, or consuming water that might be there, or changing the quality of the soils? In order for this to really be fully appreciated, we need to remember that understanding history is important to ecologists. In some way this makes ecology different as a science because these historical events are so important. So, ecological theory must encompass the past land use, the climate, the natural disturbance, and history.

Ecosystems vary across space and time, and so as we move through this course, we're shifting from a model that relied on stability as the key component, and the changes occurred infrequently, to a theory that understands that ecosystems are malleable and constantly changing. As you can imagine, that ramps up the complexity of the theories. The classic paradigms that were the building blocks of ecology encompassed words like stability, equilibrium, linearity, and control of disturbance. In fact, even the way we categorize the relationships among organisms is dynamic.

If you take a look at this image, you'll see that we have ideas like taxonomic affinity, and guilds, and functional groups. Those relationships reflect this dynamic in changing relationships. So when we think, for instance, of taxonomic affinities, we're talking about collections of organisms that are closely related to each other, like the small mammals that might be living in an ecosystem, or the birds that might be living in an ecosystem. They eat different foods, they may live in different parts of that ecosystem, but they are relatively closely related taxonomically and, hence, at their core have relatively similar needs.

Another way of bringing this idea together is to return to this idea of a guild, which I just mentioned. But these guilds can have very different organisms as part of that way of collecting them. The guilds refer to the kinds of foods that the animals need, or more generally, the kinds of resources they need. So you might be talking about organisms that rely on nectar for their food. They could be bats, or birds, or insects, but they're all relying on the same kinds of foods. So despite the fact they might be very different sizes and have different life history patterns, if there's a drought, for instance, that impacts the amount of nectar that's available, well, that impacts a broad range of organisms.

Finally, we have functional groups. Functional groups are organized around the way in which they go about solving their problems. So you might have different kinds of insects that are bunched together, and they might have very different lives, but essentially, they are solving some of the day-to-day problems of how they acquire food. They might be parasites or something like that, and so they have similar characteristics in that respect.

In the 1970s we saw this emergence of a new approach based on non-equilibrium, constant states of change. This is how we're going to view ecology, and the beginning of recognition of humans as the critical ecosystem drivers. So if disturbance is inherent to the internal dynamics of ecosystems, and ecosystems might be, then, subject to sudden, unpredictable change, what are the consequences over the long haul? What are the impacts with respect to resilience? How will this ecosystem respond to the kinds of changes?

We got a bit of an inkling into this in the early 1960s when Rachel Carson, a naturalist, and author, and scientist, wrote what became a watershed book with respect to our understanding ecosystems, and that was *Silent Spring*. In *Silent Spring*, what she did was to document a political and functional history of the use of pesticides, as well as including important segments of data. But what she really did that was so great was to begin to think of the impact with respect to its overall systems approach. So she was investigating the adverse impact of the agricultural use of pesticide, such as DDT, in the United States, and the idea that DDT can remain as residues in the ecosystem, and actually biomagnify within the food chains and food webs. We'll investigate this in more detail when we look later at biomagnifications. But it gives you an idea that here are some of the roots of systems thinking, something that's happening years before pesticide use, can reside within the ecosystem, and build up in certain species, and have downstream impacts that hadn't been considered at the time in which these pesticides were being regulated.

Emerging from these early studies like those of Rachel Carson was a shift from stability to resilience, which suggests that we should investigate resilience to a larger degree. What do we mean when we say "resilience"? We're talking about the ability of an ecosystem to undergo disturbance, and maintain its functions and controls. So, the disturbance is integral, but it's not destructive.

Another way to think of this is to think of ecological persistence. In other words, the core characteristics of the ecosystem exist over long periods of time, despite the fact that they are buffeted by constant change.

If we think about constant change, we can divide these forces of change into some of their fundamental characteristic parts, and that would be thinking about them with respect to their time course. Do they occur in short doses, or do they occur over long periods of time?

Ecologists have a new model for understanding this dynamic nature. The most important features of this model are that it includes both short and long-term events, and integrates both earth science and human impact. This model was developed jointly by the National Science Foundation and a group of ecologists who were working in long-term ecological research. It's a way of integrating not only these different kinds of forces that are at work, but again, taking a clear look at the role that humans have in the ecosystem.

This program, or this model, call the "ISSE framework," which is Integrated Science for Society in the Environment, takes a look at a variety of forces and drivers in services that allow us to really understand how ecosystems work and change. And so we can think of a series of external drivers. We can think of ecosystem services. We can think of biogeophysical forces, biological drivers, and working at different scales. So if we take a look at a basic structure of this conceptual framework, we see that there are essentially two fundamental boxes. There is a human side to this equation, and then there is the way that the ecosystem functions within the community. Linking the two are these services, the idea that healthy ecosystems provide critical services for us, and we know that to be true.

We talked in the last lectures, for instance, about barrier beaches and salt marshes that had been destroyed around urban areas as a way to facilitate development, a short-term gain at a long-term cost that resulted in a tremendous damages to those ecosystems because the services provided by salt marshes and barrier beaches, that is, absorbing storm surge related to hurricanes, was removed. The systems were unable to absorb those shocks from the storms, and so the impact of the storms was more severe. It's an example of ecosystem services. So ecosystem services link together the human side of the equation and the biotic side of the equation.

When we think about how these systems are impacted, then, we can think about them as having long-term changes and short-term changes, what we call "presses" and "pulses," and these serve as external drivers. Then there are a series of questions that we can then pose as ecologists about how the system works.

So let's take a look at some of the drivers. At the global biophysical level, we have things like ocean currents and trade winds. Ocean currents have historically been used by humans to move vessels long distances. The same with the trade winds. But not only do they do that, they also move pollutants. They also move other organisms. Global climate change and atmospheric chemistry is a huge aspect of the large scale of forces that are at work.

At the local level, the local biophysical forces, we have things like primary production, nutrient cycling, and the microclimate. These are all things we're going to look at in detail. Nutrient cycling is the way in which materials move through local ecosystems. Primary production is really a measure of how healthy your plant community is, and microclimate refers to the specific conditions that organisms are living in, in very, very small scale. So for instance, you know with respect to microclimate, one side of your house could be much warmer or colder than the other side simply as a function of which way the wind is blowing at a given time, the direction the Sun is coming from, and so forth.

There are geologic drivers to these systems. The geologic drivers include the topography and shape of the landscape, and how that landscape is changing over time. If we're talking about a particular ecosystem, the altitude that system is at, the way in which water drains, the profiles of precipitation, the so-called "hydrogeology," what is the relationship of that landscape to water? These are critical variables with respect to the nature of the ecosystem.

At the local level, we can have things like morphology, which is the local structure in which these organisms are finding themselves. It's the localized shaping of the landscape with respect to its relationship to hydrology and geomorphology. Again, it's the shape of the land and how water is related to that system.

And then there are human social forces. This is the new component that modern ecologists are bringing to the table with respect to this conversation. Power structures—how is political power distributed within a community? That has enormous impact. Are the decisions made at the local level? Are decisions made at the top level and enforced? These are significant changes with respect to the way in which ecological decision-making plays out within a community.

What about human migration? There have been tremendous movements of people around the planet.

Human events can also be broken down into short and long-term events. In addition to the length of the event, these events also have what we call "feedback loops" in the sense that one event also has an impact on other events.

Let's take an example of a system of a feedback loop, and that would be Lyme disease and gypsy moth outbreaks. Gypsy moth outbreaks and Lyme disease tend to be a Great Lakes and New England and East Coast of the United States phenomenon. Neither are particularly beneficial to humans in ecosystems, so we actually consider that a disamenity. The key species involved with both Lyme disease and gypsy moth outbreaks are white-footed mice. In the case of Lyme disease, something we're going to look at in more detail later, but just to give you a little bit of input now.

White-footed mice carry the tick that is responsible for transmitting the bacteria that causes Lyme disease to humans, so we consider the deer tick a vector. As mouse populations increase, so do the number of deer ticks. But mice are incredibly dependent on available food resources. So as mouse populations increase, they also eat more of the gypsy moth larvae, and so as gypsy moth populations increase, the mouse population increases, and that ultimately drives the gypsy moth population back down to a low level. So as mouse populations increase, the risk of Lyme disease goes up, and the risk of gypsy moth outbreaks and tree damage go down. So we end up having a negative correlation that feeds back on each other.

I said a few moments ago that there are five central questions that help us to understand the relationship of humans to ecosystems. It's these five critical questions that help us to shape our understanding of how this new model, this new conceptual framework, this ISSE framework for understanding ecological systems, helps us to push the science forward. Let's go over those questions.

The first question is, how do long-term press disturbances and short-term pulse disturbances interact to alter ecosystem structure and function? In other words, we can categorize these kinds of short and long-term impacts, but what are their interactions? Are changes in short-term impact also impacting long-term? We don't know. This is an important arena of questions that are emerging.

Second, how can the biotic structure be both a cause and consequence of ecological fluxes of energy and matter? In other words, how does the distribution of organisms as we find them within an ecosystem either cause or serve as a consequence of some of the other dynamic aspects, or changes, or fluxes that we're measuring, especially of energy and matter? We touched on this just a moment ago with the role of white-footed mice with respect to gypsy moth outbreaks and Lyme disease.

Third, how do altered ecosystem dynamics affect ecosystem services? To me, this is really the $64,000 question with respect to the future of sustainability. So as humans alter ecosystems, as humans change the way in which rivers flow, as we change the nature of soils, as we alter barrier beaches, as we plant more croplands, as we alter the drainage of soils, and so forth, how does that impact the key ecosystem services that we're depending on? One of the things we'll discuss later, in more detail, is how these ecosystem services play off. But I think you're getting the sense that we think these ecosystem services are pretty valuable. We'll put some dollar figures to them later on in the course, but the take-home message is that, you know, ecosystem services are the most economically friendly way to solve some of the most fundamental needs of the human species: clean air, clean water, and safe soils.

Question four: How do changes in vital ecosystem services feed back to alter human behavior? This is really a question of do we learn from our mistakes? In other words, if we understand that certain ecosystem services have to be maintained, and we'll have to change our behavior to maintain them, will we actually do that? That, of course, remains to be seen.

Finally, question five: Which human actions influence the frequency, magnitude, or form of press and pulse disturbance regimes across ecosystems, and what determines these human actions? Here, we're talking about this interface between human activity and the core aspects of disturbance.

Now, we've talked about the fact that disturbance can be either long or short-term in duration. Let's consider some of the long-term presses. One of the most obvious and severe is that of climate change. This is something that's taking place over centuries, and that's the typical scale at which we begin to think about long-term presses. They are forces of change that extend over a long period of time. Forest succession is the way that forests change over time in ecosystems. Coastal erosion is another example of a long-term press. Human social movements can also be important impacts that take place over a long period of time. Demographic and neighborhood changes are also critical when we add that social box to our conceptual framework of long-term presses. Immigration, movement of populations, is another key consideration of long-term presses.

Now, what about short-term pulses? The easiest way to think of these are things like catastrophic events. For example, Hurricane Katrina in 2005. We've gone back to this a couple of times, but it is so powerful in its explanatory capacity, if you will. We've learned so much from some of the mistakes that happened both during the hurricane as it played out, but also some of the ecological management decision-making that took place around the city of New Orleans, sometimes decades before that hurricane struck.

The Boxing Day Tsunami in 2004 was an earthquake followed by a huge tidal wave, or tsunami. It was centered off the west coast of Sumatra and Indonesia, and hardest hit were the coasts of Indonesia, India, and Thailand. In a relatively short period of time, only a couple of days, a quarter million people were killed. It was one of the most deadly storms that we've ever encountered in the recorded history of humanity.

Disease and invasive species is another short-term pulse that could have significant impacts within an ecosystem. A local example in North America is the Asian long-horned beetle. It was introduced as an exotic species from China, and it's dangerous because the larvae, the immature forms of this beetle, burrow into a tree in order to get their food, and while they're doing that, it girdles, or damages, the vasculature of the tree so that it can't move water and nutrients around. Some of the initial impacts we call "stem dieback," where we see dead leaves, and eventually death. It's essentially a 100% fatal disease for trees. Each adult females can lay up to 100 eggs.

In the United States, these beetles prefer maples, elms, buckeyes out in the Midwest, and willows. We believe it arrived in the United States in solid wood packing material from China. In Worcester, Massachusetts, a city in the central part of the state, 1000 trees were removed, and eventually 6000 trees are going to have to be cut down over a 64-square-mile area as a result of this infection. So it has significant social impacts as well, because this happens to be a city that in many ways identifies itself with respect to its trees, as many as 6000 mature trees are going to have to come down.

Another example is the *tamarix* genus of trees that are collectively called "salt cedars," which are invading the Grand Canyon. It turns out there are many species in this genus. There are at least eight that have been introduced into the United States, so we refer to them collectively as *tamarix*. When they are introduced into an area, they change the soil chemistry, and they crowd out the native plants. They tend to change the pH in the soil. It produces monocultures of *tamarix*, of the salt cedar, and they're water hogs. In peak growth periods they can use up to 200 gallons a day per plant. They have been colonizing western rivers as these rivers are dammed to reduce the annual flush of fresh water, and so as these river systems have been stabilized, these trees have been able to take root. They're really wreaking havoc and running amok in these western ecosystems.

Another example is the zebra mussel. Zebra mussels have become a significant problem in the Great Lakes and in areas of Canada. They're native to the Caspian Sea, and were transported to the United States by water in ballast of transoceanic vessels. They're infesting allover the Great Lakes, and also into Quebec and Ontario. Not only do they create monocultures, but they're probably the source of a deadly avian botulism infection that has periodically wreaked havoc in that part of the country.

However, it's an interesting phenomenon because their presence in these ecosystems has actually resulted in pollution reduction in some lake systems, and is also likely to increase small mouth bass populations. So it's an interesting story with the zebra mussels. It's mostly bad, but there appear to be some mitigating impacts associated with this species.

Disturbance. When we talk about things like zebra mussels and that kind of thing, we're tending to think that all kinds of disturbances are problematic, and I don't want to give that impression because, again, ecosystems are dynamic, and disturbance is actually part of the natural process. It's important to note that not all forms of disturbance are detrimental. In fact, some forms of disturbance are actually central to ecological resiliency. For example, forest fires. In typical fire-controlled ecosystems, cyclical fires are necessary for normal succession and regrowth of the forests. However, this relationship between forest fires and now human-dominated forest systems have some interesting feedback activities that are happening.

As climate change is increasing the intensity of our most severe storms, lighting and thunder are now more severe and they're causing an increase of forest fires. Secondly, human practices in these forests of suppressing fire for a long period of time, especially on federal lands, have created excessive amounts of fuel that are on the forest floor. When these fires come through now, they burn more intensely and do more damage than they would have if earlier fires had been allowed to burn. This is an example of a human-caused management dilemma because the natural disturbance regime was suppressed. The capacity of these ecosystems to buffer these events has been reduced by human activity.

The loss of wetlands is another great example. Wetlands—forests, mangroves, and barrier beaches—have increased the impact of hurricanes and typhoons. We mentioned this before. The challenge posed by these problems requires a really integrated approach to understanding ecology of human-dominated landscapes in cities.

In the next lecture, we will begin to investigate how ecosystems are shaped by the forces and drivers that humans, as a social species, provides. So until then, thank you very much.

Lecture Five
Climate and Habitat—Twin Ecological Crises

Scope:

Human population growth and urbanization are powerful drivers of land degradation and climate change. We examine the challenge of maintaining sustainability in the face of rapid demographic change.

Outline

I. Some of the most challenging aspects of our ecologies are climate change, sustainability, and urbanization.
   A. Humans are likely to be causing these problems but also have the ability to solve them.
   B. Humans are the most important factor in the future survival and resiliency of the world's ecosystem.

II. The past few hundred years have seen incredible change in the human population.
   A. At the current rate of population growth, we are adding a billion people to this planet every 20 years.
   B. Human demography is continuing to shift; we are moving toward a more urban existence.
   C. Sustainable cities are difficult to envision in relatively wealthy nations, but the challenge of urban growth is particularly daunting in developing countries.

III. Population growth and urbanization place demands on the ecosystems used to produce agriculture.
   A. The land used by humans is only about 6% of Earth's total land mass, but it includes the ecosystems that are the most rich and diverse.
   B. An additional 12% of Earth's surface has been planted with crops.
   C. A high percentage of humans live in coastal embayments, along coastal margins, or near lakes—but the very resources that attracted us to those ecosystems are not there anymore.

IV. Cities also have energy needs.
   A. Despite the high pollution load, the United States still produces about half its electricity from coal.
   B. More than 1200 square miles of biologically sensitive river areas have been buried by coal extraction methods.
   C. Because such a small portion of our energy is produced from renewable sources, we still need to produce fossil fuels (petroleum or coal).

V. We have exceeded Earth's carrying capacity, which is the number of people that can live here in a sustainable way over a long period of time.
   A. Analyses suggest that we are actually surviving on stored resources that are not being replaced.
   B. This challenges us to develop human practices that reduce degradation and exploit renewable resources.
   C. At this point, new technologies have not been completely developed, and many of them are more costly from a carbon standpoint than their payoffs.

VI. The most important human contribution to climate change is the introduction of additional greenhouse gases into the atmosphere.
   A. With regard to climate change, the most important form of greenhouse gases that humans produce is carbon dioxide.
   B. The traditional model of how greenhouse gases are produced is not accurate.
   C. The infrared frequencies reflecting back from the surface of Earth are not trapped inside the greenhouse gases; the particular frequency generated by that radiation causes the greenhouse gas molecules to resonate and generate additional energy and heat.
   D. Carbon dioxide levels have risen by more than 30% in the last 200 years, causing significant temperature increases.

VII. More than half the world's population now lives in cities.
   A. In the developed world, that figure reaches almost two-thirds.

    B. Cities provide a strategy for conservation of resources and achieving unprecedented economies of scale. Many ecologists consider cities the key to human sustainability.

    C. Public health trends show that humans who live in cities have longer life spans and better health conditions.

Suggested Reading:

Weart, *The Discovery of Global Warming*.

Wright, *Environmental Science*, chap. 11.

Questions to Consider:

1. What are the most important ways in which humans have altered the Earth's dynamic ecosystems?
2. How do excess greenhouse gasses accelerate climate change?

Lecture Five—Transcript
Climate and Habitat—Twin Ecological Crises

Hello, and welcome back. In the last lecture, we had an opportunity to think about forces and drivers, short and long-term. We were investigating the physical scale and temporal scale of changes in ecosystems. Today, we're going to take those tools and we're going to put them right in some of the most challenging aspects of our ecologies that relate to the human experience: climate change, sustainability, and urbanization. The human population growth is on the rise, and as humans, we need to understand that not only are we likely to be causing the problems, but we also have in our power the ability to solve the problems.

Think about the last time that you went on an airplane ride. If you have flown along either the eastern or the western seaboard, you will have noticed, especially if you fly at night, essentially an unbroken line of lights. If you fly on the East Coast, you can fly essentially from Portland, Maine, to Virginia and not see, really, any break in the sea of lights except when you are literally over water.

It's a rather extraordinary testimony to the kind of impact that humans have made. We really need to study this impact because it will play out as an important component to the issue of ecosystem resilience. You know, humans are the most important factor when we think about the future survivorship and resiliency of the world's ecosystem.

There's been a recent—and the only way to describe it is recent because humans go back about a million years, and really, it's only been in the past few hundred years that there has been an incredible change in the human population. It has really dramatically changed what we call the "ecological equation." It took nearly a million years for the population of humans to reach its first billion. At the current rate of population growth, we're adding a billion people to this planet every 20 years. We're already well over 6 billion and continuing to climb.

The technological nature of humans gives us this capacity to increase our footprint, increase the impact of what we're doing. Human demography is continuing to shift. We're moving toward a more urban existence. Nancy Grimm and her colleagues have been studying this phenomenon for quite some time, and if you take a look at the chart from some of her work, you can see that as we look at the total percentage of the population, beginning in 1950 and going up through the modern times, and even projecting out into 2030, you can see that the percentage of people living in rural environments has gone down over that period of time, and that the number of people living in urban areas has gone up. That's a world trend. The box situated inside is that population trend as it exists in the United States, and you can actually see that the change is more extreme. The population in the United States living in rural settings is declining dramatically, and in urban areas is growing like crazy.

This leads us to consider that the growth is essentially a developed world phenomenon, at least the shift toward cities, and we need to make it clear that that's really not the case. The data that you see in front of you now is actually both the existing population and the projected future population of some of the major cities across the world. If you look at the most populous city right now, which is Tokyo, you can see that actually its projected growth in the future, which is shown by the red histogram, actually is not changing much in comparison to its current population level. But if you look at some of the developing nations, the cities there are like Sao Paulo, Brazil, Delhi, India, and Shanghai, you see those red histograms are much taller than any of the others around them, which indicates that the population growth rates in those developing nations, in those cities, is extreme. When we think about the challenges of sustainability in developed nations, it's hard enough for us to think about sustainable cities and sustainable human footprints in relatively wealthy nations. But in those nations that are still having developing economies, the challenge of urban growth is particularly daunting. In fact, we will spend an entire lecture talking about this challenge of urban growth because it's so central to the idea of sustainability and resilience within ecosystems.

What are the most important human impacts? If we're going to think about humans as drivers in this system, we need to think about what the impacts are. We mentioned one of them: urbanization. But with every part of a city that grows, there has to be an agricultural component that helps it support the needs of individuals living within those cities. And so agriculture is as important an impact as cities. The two sometimes are not spatially connected. The food we eat may actually be grown thousands of miles away. We're actually going to look at that a little bit more in this lecture. But the two go hand-in-hand.

So urbanization creates very, very high concentrations of humans living in a particular area, but associated with those humans is the fragmented ecosystems that are necessary to produce agriculture.

As a touchstone, we think of the colonialization, or the emergence of, Western traditions in this country, beginning about the time the Pilgrims arrived in North America in the early 17th century. When the first Pilgrims would have set foot in New England, nearly 1 billion acres of forest habitat existed in what we consider the lower 48 states region. It was rather extraordinary. These were old growth forests. Most of that old growth forest is gone. There are only remnant tracts across the United States, especially on the East Coast. What remains is actually quite different in structure than was encountered by the Pilgrims.

What we do know is that the United States has somewhere in the vicinity of about 700 million acres left, so there's about three-quarters left, but it looks very different than it did at the time the Pilgrims were there.

Stuart Pimm, who is an ecologist at the University of Tennessee, estimates that at least half of the forested land in the eastern part the United States has been permanently lost, and this has resulted in the permanent extinction of a number of bird species. This is interesting to note because only about 6% of the total land mass has actually been settled or built on. So in a sense, you're saying, well, if only 6% is built on, then how is it that we can have such a large global footprint?

I think you will all agree that even without a detailed analysis, the distribution of resources as they are found spatially in an ecosystem are not equal. Humans are very sensitive to that, even those who are not scientists. So the 6% of land that's being used by humans are those ecosystems that are the most rich and diverse, and provide the most resources for people to use. So although the total land impact for habitation is about 6%, that 6% represents the most fragile and environmentally sensitive areas.

Along with about 6–7% of the land that's built up, about 12% of the Earth's surface has been planted with crops. Traditionally, when we plant crops we plant them as monocultures. Although these two combinations together don't seem like much, remember they are the most biologically diverse habitats, and that's why they were settled in the first place. And so when we look at this high percentage of humans that live in coastal embayments, or along coastal margins, or near lakes, they've made those choices because the climate is more hospitable, and the resources are more accessible.

This is somewhat ironic in the sense that the very resources that attracted us initially to those ecosystems are not there any more. So if you're from New England, and you particularly enjoy seafood, much of the seafood that you would order in traditional restaurants didn't come from New England. Even though they might be species identified with New England, they often come from distances very far away because the ecosystems themselves, the local ecosystems, have been so degraded.

Ocean ecosystems are no different with respect to the impact of humans. Halpern and colleagues have done a significant investigation into looking at the human impact of our activities along coastal margins. They've created a world map of impact, and as you investigate this map, you find that, not surprisingly, the ocean areas that are most disturbed are those that are closest to human population centers. Those areas of ocean ecosystems that have had the least amount of impact are those that are in arctic areas or other areas that are isolated from general human use.

For example, river flood plains, estuaries, and coastal margins show the highest impact because those hold the resources that we need. So the take-home point from this is that human impact is disproportionately large in comparison to the amount of land that's been settled. Examples of this include coastal cities in the United States like New York, Boston, New Orleans, Los Angeles, and San Francisco. There are high concentrations of humans living in areas that have very, very high, or at least had, very, very high biological productivity. Remember a story about New Orleans and Boston in previous lectures. Somewhere in the vicinity of 80–90% of all the wetlands that surrounded those two cities have been removed to make transportation easier or to expand the amount of space that's available for development. But the cost of doing that has been to expose those ecosystems to more severe storms caused by environmental change, but then the impact of those storms is disproportionately large because of the fact that the normal buffering systems—in this case, the salt marshes—had been removed.

As we think about cities, remember we have to support these cities. We've talked about crop land, but another way in which these cities need to be supported is through energy needs. Despite the fact that the pollution load is high and the debt we run up from an ecological perspective is staggering, we still produce about half the electricity in this country from coal. We have a voracious appetite to gather coal, and as a result, we have developed new technologies to extract coal from ecosystems, one of which is a technique called "mountain top removal." This is particularly prevalent in Appalachia, in West Virginia, and Pennsylvania. Over 1200 square miles of biologically sensitive river areas have been buried because when they try and access the coal seam, they take the top parts of the mountain off, and they have to dump the soil somewhere, so they dump it into the valleys, and those valleys, of course, become drainage areas that go into communities.

About 2200 square miles of mountaintops have been cleared and destroyed so far, and the Environmental Protection Agency estimates that by 2010, at current rates, something like 2.4 million acres of mountain systems will be destroyed by this technique. To put that into perspective, that's an amount of land the size of Delaware. This is a significant impact on the ecosystem.

Remember that in doing this, it's not as though these corporations are taking the tops off of mountains because it's something to do. They're doing it in response to an insatiable demand for energy that we have as a nation. Again, because such a small portion of our energy is produced using renewable sources, we are still bound and yoked to the need to produce fossil fuels, either in the form of petroleum or in the form of coal.

An international example of this kind of extraction takes us back to the palm oil conversation that we had earlier. We've talked about this, but again, the palm oil itself has some utility within the human domain, but the cost of doing it, the cumulative impact of its use, exceeds the geospatial size that was actually used to produce those products.

Another thing to consider is that rapid human population growth has changed our patterns of consumption and created a series of crises on the planet, the most important of which we consider this notion of sustainability. We'll talk more in detail later when we talk about how energy and those kinds of dimensions of ecosystems function.

But as an introduction, by almost any analysis that has been done, we have exceeded what we call the "carrying capacity" of the Earth, which is the number of people that can live there in a sustainable way over a long period of time. We'll detail this later, but the analyses seem to suggest that we've exceeded the carrying capacity, and we're actually surviving on stored resources that are not being replaced. Could it be any more obvious when we think about our reliance on fossil fuels? These petroleum reserves are non-renewable as we consume them and consume other resources from terrestrial and ocean environments. They're not being replaced. We either have to look to other sources or we will suffer significant population impacts.

So I say that this challenges us to develop human practices that reduce degradation and exploit renewable resources. Remember, the planet has limited resources, and populations are continuing to grow. So we talk about alternative sources of energy, those that are renewable, such as wind, solar, geothermal, and biofuels. But remember, at this point that technology has not been completely developed and, in fact, many of our strategies for developing things like biofuels are currently more costly from a carbon standpoint than their payoffs. That's because they are new technologies.

For instance, as we discussed before, if we're using palm oil as an alternative biofuel at the point at which that biofuel is consumed, there is some immediate benefit. But the biofuel in the form of palm oil is being produced in an unsustainable way in Costa Rica and other nations around the world. The combination of burning, and the combination of soil removal, and combination transport, swamps out any advantage that we're getting from using it actually as a biofuel.

Another example is clean coal. We talk about clean coal and link to new technologies, and there are ways to link clean coal, strategies to a sustainable economic model, and that would require retrofitting plants, and it's something we can move toward, but it's not something that has been fully embraced. And so we think about this challenge of coal. It is an addiction that we are locked into without immediate alternatives.

Another thing to consider is how far does your food travel before it reaches your mouth? Food production requires an enormous amount of resources. It's particularly ironic when we consider it because, for instance, if we talk about a banana, we don't actually pay the banana to grow, but we have to produce that product somewhere, and we have to transport that product to market. Especially when we talk about food production, it's very, very time sensitive. Banana production, particularly, uses all kinds of extensive amounts of resources and fossil fuels, both to grow and move it around. Typically, in banana plantations there are extensive amounts of fertilization that has to take place, and they're often very remote from the areas in which the products are consumed. And so there's an additional cost and burden associated with the production of this kind of food.

So we have a combination of explosive population growth, and consumption of fossil fuels, and what we've done is created a condition where humans are actually making permanent alterations to the climate. As extraordinary as this may seen, it's not actually unprecedented that species could modify the climate in a major way. So we've actually learned something from the past.

For example, if we go back 3½ billion years to when life first evolved, as best we know there was actually no free oxygen in the atmosphere. The oxygen was locked up in rock, and it was primitive photosynthetic organisms that were generating oxygen as a waste gas that profoundly changed the atmosphere from one that was a reducing atmosphere, without oxygen, to that which was an oxidizing atmosphere, one that has oxygen in it. In fact, the ecologists who studied some of the early paleobiological evolution suggested if oxygen had been in the atmosphere it's quite likely that life might have evolved quite differently than what we see now.

So this notion that humans can change ecosystems with their activity has created conditions that pose both challenges to us, but also help us to see possibility. Let's talk about this notion of oxygen again. We're oxygen-using organisms and, in fact, metabolism as we see it on the planet today, most of the advanced organisms are using oxygen through a process we call "oxidative phosphorylation." That's actually how we get our energy. And so our evolution is tightly linked to these huge climate change events. Looking back historically, the presence of oxygen dramatically changed the way in which life evolved over the long haul. The question is, what are the changes to the chemistry of the Earth are we creating now? What kind of scenarios will play out in the future because of those changes?

The chemistry of the ocean is profoundly impacted by the marine organisms that live there. They can alter the pH, they can alter the temperature. If we think of the great grasslands of Africa and the American Midwest, these grassland ecosystems were actually created by the foraging patterns of the deer, bison, and antelope that lived there. The fact that they have a diet of grasses prevented the growth of trees and shaped the entire ecosystem, thus altering those biogeochemical cycles that took place there. So humans are not the first species to alter the planet. However, we are doing so at an alarming rate.

Arguably, the most important contribution to climate change that humans are doing is through the introduction of additional greenhouse gases into the atmosphere. Greenhouse gases include a wide range of molecules, including methane, water vapor, nitrous oxide, ozone, and carbon dioxide. Water vapor is probably the most important greenhouse gas, and makes up about 70% of the contribution of greenhouse gases that are around the planet right now, typically in the form of clouds. However, from the standpoint of climate change, the most important form that humans produce is carbon dioxide, which is a waste gas that we produce both from metabolism and from combustion of fossil fuels. Carbon dioxide flows out of every exhaust pipe, and every smoke pipe, and out of the digestive tract of every breathing animal and human.

Now, many people argue that carbon dioxide is a natural gas or chemical and not a problem. But the anthropogenic release of carbon dioxide tips the scale with respect to the ecological balance. So greenhouse gases are actually critically important. Without greenhouse gases, we wouldn't have life on Earth as we know it, and the chemistry of greenhouse gases is complicated. But the idea here is that the additional greenhouse gas produced by humans is changing the climate enough to be detrimental.

Now, here's where it gets a little bit complicated, and you need to stay with me. We're going to come back to this a little bit later in the course as well. But the traditional model in which we understand the way in which greenhouse gases are produced tends not to be accurate. It's generally portrayed as though the gases act as a blanket and trap the heat given off from the Earth. Although that is partly correct, it really doesn't work particularly well. In order to do that, we need to understand something about the way in which solar energy interacts with the surface of the Earth.

Now, when solar energy strikes the surface of the Earth, and we can think of, say, 100% of the solar energy that strikes the surface of the Earth, some of that is going to bounce back off of clouds, some of that will be lost in the atmosphere, and some of it will actually strike the surface of the ground, heating the ground up. As the ground heats up, some of it is re-radiated back into space as it's reflected—something we call the "albedo effect" off of things like snow and ice—and some of it is absorbed and re-radiated back as heat.

Clouds play a critical role in the amount of energy that strikes the Earth. From the image that you see here, you can see that on a cloudy day a significant portion of the solar energy that would strike the Earth is actually bounced back because of cloud cover. Again, something we call the "albedo effect." At night, just the opposite is true. At night, if there is heavy cloud cover, a significant portion of the heat reflected back from the Earth is actually trapped. In fact, remember your coldest winter nights, for instance, are those that are a cloudless sky in which the energy can be released.

But here is the part about greenhouse gases. When sunlight energy strikes the surface of the Earth, it heats it up, and the heat that is released by the Earth is in the form of infrared frequencies. As those infrared frequencies reflect back away from the surface of the Earth, they have to pass through the atoms of greenhouse gas. Here is where the blanket analogy fails. Those infrared rays that are reflecting back from the surface of the Earth are not trapped inside the greenhouse gases. Instead, the particular frequency generated by that radiation causes the greenhouse gas molecules to resonate, and in doing so, they generate additional energy and heat.

As you can see from the image, these frequencies that are generated out from the surface of the Earth and go back into the atmosphere pass through these greenhouse gases. And in doing so, cause the greenhouse gases to re-radiate energy in a slightly different frequency and cause an increase in the amount of heat. So yes, there is a net gain as far as energy retained within the Earth, but it's not happening just like a thermal blanket. What's happening is stimulation of these molecules and re-radiation of heat back toward the Earth.

Global greenhouse gas emissions are increasing worldwide. Carbon dioxide levels have risen by more than 30% in the last 200 years. The International Panel on Climate Change predicts that, if left unchecked, atmospheric carbon dioxide concentrations will range as high as 970 parts per million by 2100. As a result, we know that global temperatures have increased significantly. In the last 30 years, significant warming of the Earth's ecosystem is incontrovertible. Most climate experts agree that it is the result of the accumulation of additional greenhouse gases in the atmosphere that are the result of anthropogenic activities.

Now, in later lectures we are going to investigate how this increase in global heat changes climate. Some areas become dryer, some areas become wetter, some hotter, and some even become colder.

So what does this have to do with cities? The third topic here, that we said we would talk about, as human population levels approach 7 billion, more than half the world's population now lives in cities. In the developed world, that figure reaches almost two-thirds, and the percentage is continuing to rise.

Cities provide a strategy for conservation of resources and achieving unprecedented economies of scale. Many ecologists consider cities the key to human sustainability. Cities reduce the per capita consumption of resources. We know that public health trends around the world show that humans who live in cities have longer life spans and better health conditions.

We are going to look more closely at the relationship between cities and sustainability in our final two lectures that we will be covering in this course. In our next lecture, we will investigate the social and policy models that shape the world's ecosystem. Until then, so long.

Lecture Six
Human Society as Ecological Driver

Scope:

As urbanization accelerates across the globe, human-dominated ecosystems are where the majority of people live. We investigate the nature of human impact from the local to the global level, ponder some successes and failures of the past several decades, and ask where the pathway to sustainability lies.

Outline

I. Humans are truly ecosystem engineers: Our power structures, social organization, information flow, and cultural practices profoundly change the shape of ecosystems.
   A. Internationalization—moving resources around the world—has a significant impact on both the social evolution of humans as a species and the ecology of all living things on the planet.
   B. We need to preserve ecosystem services because they can degrade to the point that social and economic functionality would decline.

II. On a global scale, some of the most important drivers are treaties, like the Kyoto Protocol or Agenda 21.

III. At the national level, we look at some of the key pieces of legislation in the United States.
   A. The Clean Air Act and Clean Water Act are examples of some very successful legislation.
   B. Prior to the signing of the Clean Air Act, the United States had relatively high levels of lead in its water, soil, and air.
   C. The Clean Air Act led to various strategies, including unleaded fuel and catalytic converters.
   D. The Superfund program, which levies a tax at the industrial level to pay for the cleanup of polluted sites, has been incredibly effective.
   E. The National Environmental Policy Act set the standard for requiring environmental impact statements.

IV. Because the United States is such a varied ecosystem, these kinds of regulations are often best handled at the state, or even local, level.
   A. The Regional Greenhouse Gas Initiative is an effort by mid-Atlantic and northeastern states to reduce greenhouse gasses through the sale of emission allowances.
   B. The local level encompasses the most accurate depiction of the ecology of the landscape.
   C. Local communities can implement growth management acts, comprehensive plans, zoning and building codes, and the state environmental protection acts.
   D. Local zoning regulations define housing and commercial density.

V. We return to the topic of urbanization.
   A. The world is becoming increasingly urbanized, rising from about 37% in 1970 to about 60% predicted in 2030.
   B. There is a positive relationship between the percentage of people living in an urban landscape and income, but this growth in urban wealth is not equitably distributed.
   C. In Asia, industrial development leading to greater productivity has driven the movement into cities.
   D. In Africa, growth has lagged, and it is more social strife and conflict driving rural inhabitants into the relative safety of cities.

VI. What is the pathway toward sustainability? The green revolution has been one of the world's great success stories.
   A. The green revolution was a transformation of agricultural technologies and research in the 1950s.
   B. One of the developments associated with this revolution has been a real understanding of how genetically engineered crops, pesticides, and food production interact.

Suggested Reading:

Diamond, *Guns, Germs, and Steel*.

Goudie, *The Human Impact on the Natural Environment*, chap. 1.

Questions to Consider:
1. What are the key human social drivers of environmental change?
2. How have human living patterns changed, and how do these changes affect global land-use patterns?

Lecture Six—Transcript
Human Society as Ecological Driver

Hello, and welcome back.

You know, traditional ecology was a science of the wilderness, but modern ecologists have embraced the human factor. Human power structures, social organization, information flow, and cultural practices can profoundly change the shape of ecosystems. Humans are truly ecosystem engineers. We began this conversation in our last lecture, but today we're going to look much more closely at some of the social organizations that impact the ecology of our planet.

Let's think about this whole challenge to begin with about internationalization. Our food, our clothing, and electronics all come from around the world, and we've embraced the notion that we gain from a cultural exchange. But from an ecological sense, not only are we thinking socially, but we are physically moving things around the planet. We are taking resources from one part of the world and sending them somewhere else, and moving resources back and forth. This has a significant impact on not only the social evolution of humans as a species, but it also has a profound effect on the ecology of all living things on the planet.

So we know that ecology has shifted from a study of the wilderness to now an additional focus on the study of humans as ecological drivers. Humans act as drivers on policy, socioeconomics, the biogeophysical world in which we live, and we'll focus, on this lecture, on the social aspects.

The reason we need to do this is because the ecosystems in which we live can reach a point where they are too degraded to perform the functions that we need from them, and the ecosystems become unstable. Remember in that multi-box conceptual framework that we looked at, the ISSE framework, ecosystem services are the key aspect that we're trying to preserve. Eventually, these services are degraded to a point that social and economic functionality can decline. The ecosystem declines, our urban systems decline, and then our system becomes unstable again.

Let's take an example of how human impact can affect landscape ecology. This takes us back a little bit to the history of the Irish potato famine, which occurred in the middle of the 19th century. The Irish potato famine was ostensibly caused by a plant disease called "the blight" that was caused by the black fungus, *phytophthora infestans*. This fungus caused the potato harvest to fail for a number of years. It resulted in a million dying of starvation, a million people emigrating in just over three years. It was a profound change to the Irish culture and to the Irish nation. The roots of this are complicated. They have deep social components and biogeophysical components, but the end result was a significant change.

Now, in contrast, in communities where potatoes had been a core crop for generations, or even centuries, in the Andes, where the potato is native, Andean farmers plant 500 varieties of potatoes in order to avoid the kind of impact that's caused when you plant only one species. The black fungus, the potato blight, wouldn't have happened if there were multiple species of potatoes that had been planted.

In this way we see an ecosystem service, such as high levels of biodiversity, acting as a buffer to the unpredictable environmental change. In fact, Andean farmers will plant these multiple different types of potatoes at all different levels along these terraces, and in doing so, defend themselves against the kind of things that devastated the Irish landscape.

So we need to think about the socioeconomic drivers, the policy drivers that have influence on ecosystems. Let's begin at the global scale as we did in the last lecture, worked our way through different scales. At the global scale some of the most important impacts are treaties, like the Kyoto Protocol, or the Agenda 21 as it was developed by the United Nations. Remember, that involved an Earth Summit that was held in Rio de Janeiro on the environment and development in 1992. It ended up producing a protocol that was signed by 178 national governments, which adopted the standards. In this were 27 core principles. It was a human centric ecology, but it was very much focused on the idea of sustainability.

It included such elements as equality of women, environmental justice, sustainable agriculture, a stakeholder status of indigenous peoples. This is worth taking a moment to talk about because, historically, when we think about environmental interventions, we think of them as coming from the top down. But in the environmental justice movement, and in the most appropriate scale in which environmental change occurs, it is the bottom up. It is community level involvement that's absolutely critical, and the

Kyoto Protocol recognized this. Now, to date, the U.S. has signed, but has declined to ratify the Kyoto Protocol. But many countries are moving forward in trying to adopt its sustainability practices.

At the national level, the United States has had, actually, a very interesting historical relationship with the environmental movement. The Clean Air Act and the Clean Water Act are actually examples of some very successful legislation. They came about as a result of a variety of rising environmental concerns, which took sort of the public stage in 1970 with the first Earth Day, a national celebration that was spearheaded by Senator Gaylord Nelson from Wisconsin. He sponsored a national day of teaching around the country on the environment and the challenges of an overpopulated human community. Emerging out of that came a series of regulatory frameworks that the U.S. uses to structure its environmental regulation. Among other things, we have national law that defines the Environmental Protection Agency's responsibility for protecting and improving the nation's air quality, and at that time there was a very focused concern on the ozone layer.

Take a look at some very interesting data with respect to lead. You know, lead, as it emerges as an air pollutant, is one of the most serious and lethal pollutants that we contend with in a highly technical society. Lead is a byproduct of combustion of many fossil fuels, and prior to the signing of the Clean Air Act we had relatively high levels of lead in the water, and soil, and in the air. But with the Clean Air Act came the intervention of various strategies, including unleaded fuel, catalytic converters, and other forms of containment of lead in the combustion process. As you can see from the histogram, if we look at lead levels before and after the implementation of these regulations, we've seen a tremendous reduction in the amount of lead in the environment in the United States, and that's really, really good news.

Another piece of really important legislation was the Comprehensive Environmental Response, Compensation Liability Act, called "CERCLA," which you probably know better as the Superfund program. It was enacted by Congress in 1980. What it did was levy a tax on petroleum and chemicals at the industrial level. In the first five years it generated about $1.6 billion that was used for the cleanup of polluted sites. The Act was renewed in 1986, and a trust fund has now nearly $10 billion in assets that can be used for cleanup. The Superfund program has been incredibly effective in trying to go after the most difficult of pollution issues.

There's also the National Environmental Policy Act, or the so-called "NEPA," which was signed into law in 1970. It set the standard for requiring environmental impact statements. For those of us who are familiar with projects as they go forward now in the country, we expect that we would be filing an environmental impact statement that would assess the potential damage that we might cause in any ecosystem with respect to building roads or constructing buildings, something like that. But prior to the 1970s, those weren't even filed. And I think most importantly, they became the model for the state environmental programs. Because the United States is such a varied ecosystem, these kinds of regulations are often best handled at the state, or even local, levels because of the differences that you see in the ecological characters of the particular states, and communities, and towns.

Another, I think, really important regional intervention has been the so-called "RGGI," the Regional Greenhouse Gas Initiative. It's an effort by mid-Atlantic and Northeastern states to create mandatory, market-based effort to reduce greenhouse gasses through the sale of emission allowances. In this case, the proceeds from these allowances are used to spur clean renewable energy research and development. The goal seems modest, but it's pretty significant. It's to reduce CO_2 emissions by 10% by 2018, which involves not only curbing the increase in CO_2, but managing to reduce it to 10% below the levels that we see now.

I would also argue that one of the most important scales of intervention, and again, this is coming from an ecologist who looks at these challenges and different scales, are the impacts that can happen at the local level. Remember, the local level is going to encompass the most accurate depiction of the ecology of the landscape as it sits. And so local communities can implement growth management acts, comprehensive plans, zoning and building codes, and of course the state environmental protection acts. Local zoning boards can have a huge impact on things like whether or not big box stores or mall developments happen in the community, or even lot sizes.

Zoning regulations are interesting because they define housing and commercial density, and this is one of the challenges that we face when we begin to think about how ecology maps onto what are essentially political attempts at intervention. So let's say we have a community that is having a challenge by extensive growth that is putting a significant burden on the ecosystem services within the community. Well, some of these communities will respond to increasing population by zoning for bigger lots. In some communities this goes up to three acres or even more. While that does limit the number of individuals who can move into the community, it doesn't actually solve the problem of fragmentation because by doing a series of zoned lots, you still end up breaking up the ecological integrity of the community. I think worse, when you begin to implement these kinds of zoning regulations, especially as the lots become bigger and bigger and bigger, you're actually pushing away a whole cross-section of the community that can no longer actually afford to live in the town that they might be working in.

So an ecologist would come to the table and say, "Perhaps we need to think differently about local zoning." One strategy toward this is to think about something we call "cluster zoning," where we put our residences in relatively close contact with one another, but it allows for expanded amounts of green space away from those dwellings. Some communities are approaching this. So instead of taking ten houses and putting them each on three-acre lots, let's take these ten houses and put them on a smaller collection of lots and let's preserve the other 25 acres for access by members of that community.

Now, local zoning ordinances can actually force developers to acquire existing parcels. This is a very interesting response to environmental need. Remember we talked about this notion of the tragedy of the commons a couple of lectures ago? Local zoning ordinances fall into that category because if you're a developer, you might consider rebuilding an area that has already been used. But what you don't know about the history of that area is in response to, what are the impacts the previous owners have done? Is there oil in the ground? Is there hazardous material on that site? Is there a legacy of human activities at that site that's going to come back to bite you as a developer later on? And so what happens, because of that fear of the unknown, developers almost always historically have preferred to go to new sites and put up new buildings, as opposed to reusing old ones. And so what we've seen when that kind of response is used, or that kind of building strategy is used, we see what we consider classical sprawl, where the inside of cities or inside of towns begin to decline. The buildup goes to the outside, and you end up with large numbers of buildings that are being unused.

Now, redevelopment authorities, like the one that we have in Boston, can actually develop strategies that make it more appealing for developers to reuse existing sites. They can do a number of things. They can expedite plan reviews, for instance, where developers reuse old buildings, but bring them up to modern certification with respect to sustainability. And federal and state regulations can limit the liability of new owners with respect to what has happened in the past before they owned the land. In that way, you have taken away some of the disincentives to moving into existing buildings. This is a way of balancing that public-versus-private cost benefit analysis that developers have to use.

Let's think about global socio-economic drivers. This is on the very largest of scales. Political regimes, trade policies, exchange rates, world markets, commodity prices, all of these have profound impacts on world environmental conditions. We talked about the palm oil plantations before, and we discussed how dramatically they alter ecosystems, especially tropical landscapes, around the world. I want to talk about another challenge to global ecosystems, essentially on the same scale as the palm oil plantations, and that's worldwide shrimp farming. Now, Americans have become particularly fond, as have the Japanese and European markets, of having shrimp as a regular part of our diet. It's become a $9 billion annual industry. Almost 1.7 million tons are harvested annually.

Remember we started out this conversation talking about the fact that not only are there cultural, but there is physical distance between the resources we use and where they are produced. Shrimp farms in Asia produce 75% of the shrimp that is consumed annually. The other 25% mostly come from Latin America, and of that, mostly from Brazil. The whole shrimp farming industry has moved from small local practices to much more concentrated, sort of ultra-sized farming, mega-farms. This is now a global industry, and essentially, only two species of shrimp are grown: the Pacific white shrimp and the tiger prawn. So almost like the story about the potatoes, we've gone from having lots of different varieties of shrimp down to these monocultures, and these monocultures are susceptible to disease, such as white spot syndrome and yellowhead disease, and when the monocultures become diseased it wipes out entire farms. So the response of the people who runs these farms is continued rapid expansion and movement of the farms, leaving behind a legacy of degraded landscape.

Because wetland systems are the preferred areas in which these shrimp are grown, as much as 10% of the world's mangrove ecosystems have been lost to make way for shrimp farms. Like any concentrated farming—and we have our own terrestrial examples in the United States where we have intensive cattle and hog farming—concentrated feed and waste have devastating local impacts on ecosystems.

So, in essence, here we are in developed nations, consuming our shrimp for dinner, and we get the benefit of the shrimp, but the ecological burden is transferred to developing nations, who are seeing this as a short-term way to enhance their economy. But in the long term it's having significant detrimental effects, and it's causing the collapse of local ecosystems.

Now, in response to this, at the end of the last century, in 1999, the World Wildlife Fund and the World Bank partnered with a series of sustainable aquaculture experts from these various countries, and they have developed what they call a "Healthy Practices Model." It's still in the early stages of development, but it does look like it may be an initial pathway toward sustainability on that front.

As you can see, we have two models here that we've talked about that are very challenging, and they're driven by consumer demand in developed nations: shrimp farming and palm oil production.

We need to return to a conversation about urbanization because the development of cities has caused a demographic shift, and in this demographic shift, remember, if people are living in high densities in urban settings, they can't, for the most part, grow much of their own food. They're going to rely on this from other sources. And as we mentioned before, the world is becoming increasingly urbanized, rising from about 37% in 1970 to about 60% predicted in 2030 worldwide. The most important part here is that urban areas will see 95% of the world's population growth through the next 30 years.

World income has grown in response to urbanization. Work by David Bloom and colleagues at Harvard School of Public Health have looked closely at the relationship between real income per capita as members of a population, and the percentage of the population living in urban areas. If you take a look at the figure, you can see there's a pretty extraordinary relationship between income and percentage of the population living in urban areas. As the population of individuals living in urban areas goes up in any particular ecosystem, in this case it's politically defined as nation, you see that income goes up as well. So there's a positive relationship between percentage of people living in an urban landscape and the income. This engine of economic growth drives many of the fastest growing cities.

In the United States, Las Vegas is one of our fastest growing cities. It is growing about as fast as it can. There's a tidal wave of sprawl that comes with these developing areas. And although in cities like Las Vegas the growth is driven by economic expansion, it turns out that Bloom and his colleagues have discovered that this growth in urban wealth is not equitably distributed.

If we take a look at another graph from his paper that appeared recently in *Science*, you can see that they have tracked the population percentage that's living in urban areas from 1960–2000 and beyond, and then they also looked at the change in real income. They compared both Africa and Asia, which, as you can see from the top lines, have very similar growth patterns. Both have grown from about 20% urban population in the 1960s to about 35% population living in urban areas by the year 2000.

But look at the tremendous difference in the growth as far as real income. In Asia, the growth in real income has been extraordinary. In Africa, the growth has been slow or, in some instances, even declined. This represents the different social drivers, again, the ecology of the system, different social drivers leading to this increase in urbanization.

In Asia, it's industrial development leading to greater productivity that's driving the movement into cities. Sadly, in Africa, it's more related to social strife and conflict, driving rural inhabitants into the relative safety of the urban landscape. So it's a migration driven by fear. And as you might predict, then, the response is not an increase in economic earning potential. It's simply moving from rural areas to urban areas. So this complexity of outcomes is a signal that urbanization is not always going to lead to enhanced quality of life.

Now, world governance, and government structures, and social organizations are often going to determine how resources are used. World consumption and production of energy is a very an uneasy balance that results in conflict and unexpected alliances among suppliers and consumers. Again, Hardin's tragedy model fits very well here. We think about the areas of the world that have the highest amounts of conflict. They tend to have that conflict because they have areas that are the most important sources of resource.

China and India are growing populations, and their use of resources and effect on the planet is rapidly increasing. In fact, in 2006, China surpassed the United States in annual greenhouse gas production. In that year the U.S. produced about 5.8 billion tons. China produced 6.2 billion tons. I would point out, however, that China has a billion people, and we have about a third of that, and so the per-capita use is still extraordinarily high in the United States. In China, the increase came primarily from new coal-fired electrical plants and from the massive production of cement that they need to expand their urban infrastructure.

Now, some would argue that this shows the fallacy of global treaties such as the Kyoto Protocol. So here we've had the Kyoto Protocol in place for a while, but we still have countries like India and China growing and doing so in ways that increase pollution. But others would suggest that this pattern actually reinforces the disparity between wealthy and developing nations. I would argue that these patterns exist as a function of the conditions in which these nations finds themselves.

So, then, what is the way out of this? What is the pathway toward sustainability? I think one example would be to look at agriculture and the Green Revolution, which I think has been another one of the world's great success stories.

Remember, the Industrial Revolution of 18th and 19th centuries gave us tremendous technical capacity. It led to the development of machines and an entirely new vision of what it means for humans to be involved with labor. There has been a switch toward technology for production of goods and services. There's been hyper-concentration of resources in urban production centers. And, initially, people living in urban areas suffered what we call an "urban penalty," and that was related to how dirty and pollution-filled, and disease-ridden cities had become. They no longer have to suffer that burden, but initially they did.

After the birth of urban public health movement, the rising standard of living and birth of the middle class really changed cities so that we can see, as we did in Bloom's work, that the consistent model, or the overall model, is that as urban populations increase, so does real income.

The Green Revolution followed the Industrial Revolution, and it was a transformation of agricultural technologies and research in the 1950s. We're going to focus in more detail on the agricultural revolution, but I want to take a look at the agricultural revolution as a social driver.

Now, Mexico was a major leader in the development of the Green Revolution. They created new ways for wheat production to help feed their rapidly growing population. In 1943, Mexico imported half its wheat. By 1956, Mexico was self-sufficient as far as wheat production, and by 1964, Mexico exported half a million tons of wheat.

The Agricultural Revolution has also led to the admittedly more controversial development of genetically engineered crops, and pesticides, and an increase in our world's food production, which has had mixed results. But the extraordinary thing is that we are actually able to feed our population.

One of the developments associated with the Agricultural Revolution has been an emergence of a real understanding of how these systems interact. And so now, in fact, in sustainable agricultural systems we see patterns of use, such as integrated pest management, where the appropriate interplanting of multiple species of plants can actually reduce the need for pesticides and fertilizers.

In addition, we also see the use of strategies such as crop rotation, which is a way for farmers to rest the land. It is actually an old practice made reference to in biblical texts, the idea that we won't grow the same crop continuously year after year after year within an ecosystem. We will continue to rotate the crops, some of which actually put nourishment back into the soil as they're turned over. It's actually an old idea, one that was brought forward by Aldo Leopold in his *Notion of the Land Epic*, but one that's finally making its mark on an international scale.

In the next lecture, we're going to take this idea of food and consider how energy moves through ecosystems from a scientific perspective, the role that organisms play in moving the energy through what scientists call the "trophic cascade." So until then, farewell, and thank you.

Lecture Seven
Movement of Energy through Living Systems

Scope:
The movement of energy through ecosystems can be studied via trophic pyramids and food webs. Humans and other animals have many food options, and their choice of where on the food chain to eat greatly affects energy use. We discuss some of the ongoing research about interaction within food webs.

Outline

I. We examine energy flow through ecosystems.
   A. All energy in ecosystems is ultimately derived from sunlight.
   B. Energy flow in ecosystems is unidirectional, traveling through what we call trophic pyramids.
   C. The rules of energy transfer dictate the abundance of species and the population density of all organisms in an ecosystem.

II. We look at the trophic pyramid of energy use.
   A. Energy from the Sun is captured by organisms we call producers, which include algae and plants.
   B. The organisms that feed directly on plants (herbivores) are called primary consumers.
   C. The carnivores that eat the herbivores are called secondary consumers.
   D. Carnivores that eat the secondary consumers are called tertiary consumers.
   E. There are also groups of organisms (like fungi and prokaryotes) that we call decomposers; they help to recycle the materials between each trophic level.

III. The trophic efficiency between each level of the pyramid is quite low.
   A. Most of the energy is used to run the metabolisms of the organisms.
   B. The loss of that energy within a food chain is reflected in the shape of the energy pyramid, which tends to be very wide at the base and narrow at the top.

C. Humans can eat a carnivorous or vegetarian diet; depending on where we eat within the food chain, we use energy quite differently.

IV. The terrestrial food web is complex.
   A. In a diagram of the web, the arrows show the direction in which the calories (energy) are traveling.
   B. Animals have many diet choices.
   C. Those organisms that can feed at different trophic levels generally have the highest success in surviving around humans.
   D. There is a lot of controversy over whether increased complexity is beneficial or harmful to ecosystems.

V. Current research investigates the strength and interactions within a food web.
   A. Removal experiments on the sea star led biodiversity to plummet, leading the sea star to be considered a keystone species.
   B. A study of wolves, moose, and tree growth on Isle Royale revealed that the herbivores were responsible for regulating tree growth.
   C. Recent studies in Yellowstone National Park have shown the impact of top-down control on plant communities along rivers and streams.

VI. Another result of the movement of energy through systems is the emergence of ocean dead zones as a result of human disturbance.
   A. There are nearly 400 such dead zones distributed around the world, covering nearly 95,000 square miles.
   B. They develop because of eutrophication as a result of nutrient loading, usually from fertilizer runoff.
   C. The good news is that ecosystems are resilient—this whole process is reversible.

Suggested Reading:

Cain, Bowman, and Hacker, *Ecology*, chap. 20.

Wright, *Environmental Science*, chap. 3.

Questions to Consider:

1. From a general perspective, how does energy interact with natural ecosystems?
2. What are ecological food web models, and how do they help explain the movement of energy in an ecosystem?

Lecture Seven—Transcript
Movement of Energy through Living Systems

Welcome back. From here through Lecture Twenty, we're going to investigate specific forces and drivers that cause ecosystems to change over time. We're going to start with energy. What we're going to do is, in the first of these sort of coupled lectures, we'll talk about the system as it operates, and then in the second lecture we'll look at the human impact on those systems.

Now, all energy in ecosystems is ultimately derived from sunlight. Energy flow in ecosystems is unidirectional. The ecological users of energy are organized into what we call "trophic pyramids." The rules of energy transfer dictate the abundance of species and the population density of all organisms in an ecosystem. The energy gets distributed in something we call "complex food webs."

To begin with, I want to introduce a stuffed squirrel here. It's actually a study skin. It's prepared for study, so it's stretched out. This is *sciurus caroliensis*, the gray squirrel. It's very common throughout the Eastern and Midwestern areas of the country. It's what we call a "scatter hoarder," which means that throughout the summer it collects thousands of food items and caches them each year, so it has food to eat during the winter and early spring.

In order to do this, this animal has extraordinary spatial memory. It literally remembers thousands of places that it stores food. It's a critical species in the food web because it both consumes many forms of seeds and nuts, but it's also eaten by many predators. In fact, it doesn't actually remember all the places it stores food and, as such, it actually produces food resources for a variety of organisms that break into the squirrel's cache and gain access to those resources.

Energy flow in ecosystems is unidirectional, and it's governed by models of thermodynamics. What that means is that all the energy that we use ultimately traces back to the Sun. Energy arrives from the Sun, and some of it is converted to chemical energy. That energy can be used by other living organisms. All of the energy in ecosystems is ultimately degraded to heat and lost to the atmosphere.

Now, it's important to note, as organisms, we can conserve energy but we can't reuse it. So the notion of reuse and recycling is an important idea when it comes to the materials in ecosystems, that energy flow is essentially unidirectional. These relationships differ significantly between materials and energy. Materials, of course, do cycle. We will investigate in a future lecture the whole notion of material cycling.

We need to begin by investigating this energy in more detail, so I direct your attention to a figure around the movement of energy through ecosystems that I think will be helpful. In reviewing, remember, the Sun is the ultimate source of all energy in living systems, but only about one-tenth of 1% of that is actually converted by photosynthetic organisms, and we call that the "gross primary productivity." In other words, the amount of energy that's actually captured by photosynthetic organisms.

Of that, only about 1% of the energy captured is actually found in the tissues of plants and photosynthetic algae, which is the stuff that consumers and other organisms in the system can actually eat. So the rest of that energy is actually used for growth, and repair, and other activities within these photosynthetic organisms. And so about 1% of that energy is actually transferred, and we call that the "net primary productivity."

So, as you see, and we'll look through this in more detail, energy from the Sun, captured by organisms that we call "producers," which are traditionally what we think of as being plants, and these are the organisms that harness the Sun's energy for food production. We call them the "producer group," and it's algae, plants, and some photosynthetic bacteria.

Producers make their own food through the process of photosynthesis. It's an extraordinary process where the particular energies of light excite the particularly energy-absorbent molecules within the photosynthetic organism's metabolism, and those energies are actually captured in chemical bonds. That locks them into availability not only to the photosynthetic plants, but to the other organisms as well. Light energy is absorbed, and carbon dioxide and water are converted into carbohydrates, or food, and oxygen as a waste gas.

Heterotrophs are consumers that depend directly on the photosynthetic output of plants as a source of energy. We call the organisms that produce their own energy "autotrophs," self feeders. Heterotrophs, other feeders, means that those are organisms that are relying on plants to provide tissues that they can consume.

The consumer level within trophic systems begins with organisms that eat the producers, such as herbivores, which eat plants directly. All animals, including humans, require the photosynthetic activity of the producer group to provide the tissues that we can live on. Humans cannot synthesize organic molecules from sunlight and carbon dioxide. The most we get from sunlight is a sunburn. We do actually transform some of the chemicals in our bodies in the presence of ultraviolet radiation, especially active forms of Vitamin D, but that's a different story.

We call the organisms that feed directly on plants "primary consumers," and they are at a trophic level just above the producers. The secondary consumers are those carnivores that eat the herbivores. Carnivores that eat the secondary consumers are called "tertiary consumers." And then within this system there are also groups of organisms that we call "decomposers" that help to recycle the materials between each trophic level, and that recycling of the materials, the breaking it down into simpler forms, actually allows energy to move between trophic levels more efficiently.

These decomposers consume nonliving organic material that we call "detritus." This trophic level plays a key role in recycling materials back into the biotic realm. Key examples of these are fungi and prokaryotes. These are the main categories of decomposers. If you're walking in a traditional woodland area it's not unusual to see mushrooms growing on wood that has landed on the forest floor or trees that have fallen over. What you're seeing, actually, is a process by which these materials are released back into the system. These organic materials are typically broken down by special enzymes that these decomposer organisms have. What they do is they actually eat the breakdown products. So they're not actually eating the wood directly; they're breaking the products down and eating the products that are produced from that.

We can envision this whole relationship as a trophic pyramid. In this kind of system, the bottom of the pyramid is composed of the producers, and these organisms, by definition, are going to have the most amount of biomass. In fact, the total weight of all living organisms in a system is what we call the "biomass."

Energy moves from one trophic level to another through an ecosystem, from primary producer to primary consumer, to secondary consumer, to the last group of consumers, or apex of the pyramid. The least amount of energy or biomass is found at the apex. And that's because of a fundamental aspect of the thermodynamics of ecosystems. You see, energy is lost between each level of the pyramid. Think about this for a moment. Organisms that are consuming energy are not doing this to provide food for other organisms in an ecosystem. They're acquiring and metabolizing chemicals that provide energy so that they can complete their own life cycles. Part of that is for growth, and part of that is for repair, and movement, and reproduction, and only a portion of the energy consumed actually gets integrated into the tissues and locked up as chemicals that can be broken down by other organisms. So, inevitably, the trophic efficiency, the transfer of energy, from one level of the ecosystem to the next is actually quite low because most of that energy is actually used to run the metabolism of organisms.

The production transfer from one trophic level to the next is only about 10%. The loss of that energy within a food chain is reflected in the shape of the energy pyramid, which tends to be very wide at the base and narrow at the top. As I tell my students, well, that's why, in a very basic sense, that when you go out into the ecosystems of New England, there are many more blades of grass than there are going to be coyotes because they're at very, very different parts of this trophic pyramid, and the amount of energy that's available to the grasses is much, much larger than the energy that's actually available to organisms that are sitting at the top of this food pyramid. This inefficiency of energy transfer has important implications, in fact, for human food choice and sustainability.

If we take a look at this relationship between where humans choose to feed on the food in this trophic pyramid, it's actually quite interesting. Humans, because we're omnivorous, can have our calories come from a predominantly carnivorous diet or predominantly vegetarian diet. And depending on where we eat within the food chain, we end up using energy quite differently.

For instance, it takes about 1000 calories of cereal grain to produce 100 calories of beef. Now, those cereal grains could be consumed by humans directly, and they would provide 10 100-calorie meals, as opposed to one 100-calorie piece of beef. We will look at this in much more detail later on in the lecture series because the story about the distribution of energy through ecosystems and the balance in vegetarianism versus carnivorous diets is really quite interesting.

In fact, it gets delightfully complicated, if you will, because as we grow different kinds of foods the energy transfer can actually be quite different.

But in order for us to understand this in some detail, we need to think about the movement of energy as it works within the system, and there we're really talking about a food web. Here, the trophic pyramid is the simplest way to examine energy flow in a food chain. However, it's often much more intertwined and complex than it is linear. And so if we take a look at even a simplified food web—and by the way, these food webs are organized, and if you take a look at the diagram you see that they're organized so that the arrows actually show the direction in which the calories are traveling. It's a little bit backward, counter-intuitive, because you might think, oh, the arrow should be pointed in the direction of this animal eats that thing. Well, instead, the arrows are positioned so that you see which way the energy is flowing.

So if you take a look at a traditional simplified terrestrial food web, what you see is that you have certain animals that are consuming plants directly and the arrows move from the photosynthetic producers to the first level of consumers. So you might see deer that are consuming a shrubbery, and then you see that the calories from the deer flow to animals that are carnivorous predators. All of that is true, but what happens, as you can imagine in ecosystems is that animals have many, many diet choices. Some animals can feed as carnivores and some feed as herbivores. In fact, as we'll see in human-dominated ecosystems, those organisms that can display the ability to feed at different trophic levels generally have the highest success in surviving around humans. Each herbivore, for instance, often eats many different plants, and is preyed on by many different predators who fall into different levels of consumption. So in nature, actual food webs are really quite complex.

Early researchers in ecology, such as Eugene Odum, believed that more complexity meant more stability. In other words, as ecosystems became more complex, there were more pieces to them and, as such, the additional pieces provided sort of a safety net, if you will, of how these systems operate. In fact, many of the early ecologists considered these additional complexities to be very, very important. In fact, coming out of that were such models as the rivet model of the structure of ecological communities where we envisioned the structure of an airplane, for instance, with many rivets that hold it together. We could consider in an ecosystem that each species was a different rivet in the ecosystem, and as you began to pull rivets out the system became more unstable and, in fact, ultimately the plane would crash. This was a metaphor that was built by Paul Ehrlich, in fact, to think about the structure of populations.

Robert May, another very powerful mathematical ecologist with respect to the ideas that he provided, shocked the scientific community that studied food webs by developing some rather sophisticated simulations that suggested that complexity can actually lead to collapse and extinction. The additional complexity, in some respects, makes the system that much more susceptible to diversion.

Current research attempts to investigate the strength and the interactions within a food web. We're trying to get to the bottom of this because as we mentioned before, one of the delightful aspects of ecology is that there's lots of controversy. We're trying to look at a systemic science, and so lots of theories are being tested.

For example, the sea star *Pisaster* has a strong effect on a variety of its prey species, mostly bivalves, mollusks, and barnacles. When Robert Paine from the University Washington did removal experiments of the sea star, the biodiversity plummeted as the top-down predation impacts were reduced. Mussels ended up dominating the benthic landscape, and Paine considered the sea star a keystone species because of its strong effect on the ecosystem. This idea of keystone species is something we'll investigate in more detail. Humans, for instance, are keystone species in urban ecosystems.

Now, these impacts can be direct and indirect. Seas stars directly impact mussels and barnacles, in this case negatively, by eating them. But for barnacles, the impact is actually positive indirectly by changing the competition dynamics that barnacles have with mussels. Sea stars eat mussels, which reduces the competition barnacles have for space on the ocean floor. So in this case, the presence of the sea star reduces the competition and so it allows for a greater degree of biodiversity. Most importantly, it allows for a greater degree of evenness in the distribution of species. We investigate this idea in more detail in other lectures, but the notion of diversity is not only important, but it's important that there by relatively similar representation of different species within an ecosystem. We call that idea "evenness."

Now, another classic study that has looked at the distribution of species on feeding and ecosystem function is that of the study of wolves, and moose, and tree growth on Isle Royale. We actually refer to the study a number of times because it's such an important contribution to our understanding of how ecosystems function.

Now, this was an investigation by Brian McLaren and his colleagues from the Michigan Technical University. These findings were published in *Science* back in 1994, but they still serve as a model ecosystem for us to compare other work to and to consider how our theories will map onto the data that were recovered from the study. It's really an extraordinary example of what I consider to be theory in practice.

So what they were really doing was studying the growth of the balsam fir, which is a key tree species in that ecosystem. What was revealed was that an intense relationship between the herbivores and the trees exists, and that the herbivores were actually responsible for regulating the tree growth.

Let me step back and discuss this for just a moment. If you were a young tree, you're a seedling and you're coming out of the ground, you've got a number of things that are going on. First of all, you're trying to compete for access to sunlight, so you need light gaps. You need the sunlight to penetrate to the floor of the forest, which by the way, is one of the reasons it's so critical that organisms of all kinds disperse because when you're growing up out of the floor, who is the individual that's actually blocking your sunlight? It's your own parent. In animal systems it's the same way. If you're growing up next to your parent and you have a territory adjacent to your parent, you're both competing for the same food resources. So dispersal is something we'll discuss in the lectures in much more detail. Dispersal is actually a key factor for not only individual fitness, but population survival.

So you have these young trees that are fighting for light gaps, but they're also trying to get resources out of the ground that they need, micronutrients. And so there is a bottom-up control for how fast they can grow because it's a function of resource acquisition. But then at the same time they're also struggling to avoid being eaten by predators, which is a top-down control.

It turns out at Isle Royale the top-down control was the most important aspect, and the species that was responsible for that top-down control was moose. Moose love the young plants because they're succulent, they haven't produced all of their secondary defensive compounds yet, and they're easily digested. So plant growth was regulated by population fluctuations in the predators and herbivores. The plants increased their primary productivity only when they were released from consumption of their tissues by moose. It turns out that moose populations were limited by predation from wolves. Let's take a look at these data. They're really quite fascinating. I know it's a bit of a complicated chart, but follow me on this because there are really a number of parts here.

First of all, at the top of the chart you see the distribution of wolves over the period of the study. You can see first that the number of wolves begins to drop, and then there's a secondary climb in the population, and then a fairly severe drop, and then recovery, and so forth.

If you notice, the population of moose, which is in the second graph from the top, follows in a cycle with respect to the wolves. So as the wolf population is declining, the predator pressure on the moose is relieved, and the number of moose increase. Then as the population of wolves increases, then you see that the population of moose decrease.

Here is where it gets very interesting, though. In the following two graphs you see the balsam fir ring width in two different sample plots, which is a measure of how fast they're growing. Notice that as the population of moose increases, the growth rates of the firs decline in both of the sample plots. And as the population of the moose decline, the trees recover. So you see this extraordinary trophic cascade where, ultimately, the growth of the firs is linked directly to the predator pressure of the wolves, which are all interconnected. So this is a clear example of what we call "top-down control." Ultimately the wolves controlled moose, which in turn controlled the growth of the firs.

Recent studies in Yellowstone Park have shown the impact of top-down control on the plant communities along the rivers and streams. The reintroduction of wolves forced the elk back onto the mountain slopes, and this release of herbivory pressure on willows along the rivers allowed them to rebound. We look at this in more detail in a future lecture.

Now, another aspect with respect to the movement of energy through these systems has been the emergence of ocean dead zones as a result of human disturbance. A study by Robert Diaz at the Virginia Institute of Marine Science, and Rutger Rosenberg from Gottenberg University, investigated this phenomenon of oceanic dead zones. It turns out that there are nearly 400 such dead zones distributed around the world. You can see them from the map here. You can see their distribution. Overall, this covers nearly 95,000 square miles. They develop because of eutrophication as a result of nutrient loading, usually from fertilizers. What happens is this rapid and explosive growth that is a result, essentially, of human pollution results in hypoxic waters. In other words, waters starved of oxygen that end up having less than 2 parts per million of available oxygen. Normal freshwater is somewhere in the vicinity of 10–12 parts per million, and even more in some areas. Think about trout, which you find in bubbling cold brooks. That's because cold water holds more oxygen, and the cold riffling water is additionally oxygen-rich because it's constantly exposing surface area so that oxygen can diffuse.

Ocean systems are somewhat less rich in oxygen, typically between six and nine parts per million. The largest dead zone, which is in the Caspian Sea, itself is 25,000 square miles. This has resulted from high levels of fertilizer runoff. Now, this process is conceptualized in more detail in the following diagram. You can see that the diagram shows the distribution of oxygen as a function of energy to mobile predators as a sort of a surrogate measure of ecosystem health. So in the green zone we have a normal amount of oxygen. Energy flows as carbon from benthic community to the mobile predators, the fish and free-swimming invertebrates.

In the orange zone, an influx of nutrients, typically from runoff, causes a short pulse in productivity, followed by decline. That decline is the result of the oxygen consumption during this burst of development. Sometimes these are seasonal events, and they can become more severe if there's human pollution. Ultimately, the systems enter a red zone, which is anoxic, no oxygen, and all available carbon is going to the anaerobic benthic microbes.

Now the good news is—and when you look at a chart like this it's pretty frightening, but the good news is—ecosystems are resilient. In fact, this whole process is reversible. The Black Sea has one of the world's largest dead zones, about 15,000 square miles. Between

1973 and 1990 this dead zone became gigantic, functionally because of the fertilizer runoff that was occurring. That was a result of fertilizer subsides that the Soviet Union was providing. And during the breakup of the Soviet Union, those subsides ended, and nutrient loading dropped to only about a quarter of the previous levels during those subsidies. So these systems began to recover, and overall, the anoxic conditions can be reversed. So it's one of the sort of exciting pieces of good news.

In closing, I just want to think a little bit more, as we transition into our next lecture, about some of these aspects of human-dominated ecosystems. When we think of urban areas, the movement of energy becomes more complicated for organisms that live there. In many instances, the fitness of those organisms is linked to their ability to live or survive, if you will, at various levels on the food chain. Urban food webs are dynamic because humans contribute to those food webs. So the successful organisms in urban areas are those that can change their diets as the fluctuating and dynamic availability of human food resources reveals itself.

In our next lecture, we're going to look more closely at the role of human energy use on global ecosystems. Until then, thank you very much.

Lecture Eight
Humans as Energy Consumers

Scope:

In this lecture, we focus on the ecological reality of thermodynamics and its implications for human survival. Humans have a voracious demand for energy, and our patterns of energy consumption produce enormous short- and long-term challenges to ecological sustainability.

Outline

I. Human demand for energy is increasing, and its acquisition has significant consequences.
   A. Somewhere in the vicinity of 85 million barrels of oil are consumed worldwide each day—19 million of them in the United States.
   B. Each year, 6 billion tons of coal are used worldwide—more than 1 billion of them in the United States.
   C. Since the late 1960s and early 1970s, the United States' ability to meet its own demand has fallen: Our production levels have tapered off, but our consumption has continued to grow.
   D. In 1950, average energy consumption per person in the United States was a little more than 200 million BTUs each year.
   E. Current consumption is around 330 BTUs per person—a staggering amount.

II. Linked to this demand for energy is a voracious demand for water, and habitat destruction from loss of freshwater supplies is extraordinary. We look at the example of the Three Gorges Dam in China.
   A. To make way for this dam, the Chinese government displaced about 1.2 million people.
   B. Its construction has spawned landslides and increased the risk of earthquakes.
   C. The dam will, however, offset about 50 million tons of raw coal per year that China would otherwise use to produce electricity.

- **D.** The dam will also become the world's single largest producer of renewable energy.

III. Patterns of energy consumption by humans increase greenhouse gases due to our consumption of fossil fuels.
- **A.** Average surface temperatures around the globe have increased about 0.6°C during the past hundred years.
- **B.** There has been a 10% reduction in snow cover ice since the late 1960s.
- **C.** This significantly reduces Earth's albedo effect, a measure of the reflectance of a surface with respect to sunlight.

IV. There are significant pollution impacts from petroleum, and they can be devastating to marine and terrestrial ecosystems.
- **A.** The effects of oil spills last for decades.
- **B.** The Exxon *Valdez* spill in Alaska in 1989 caused a near-total collapse of the local sea populations of clams, herring, and seals, which also caused 2 fishing corporations to go bankrupt.
- **C.** Reliance on petroleum has increased global political tensions.
- **D.** Extraction machinery and pipes must enter and cross many national borders, allowing small nations to exert tremendous influence by hindering petroleum deliveries.

V. Pollutants released into the atmosphere can travel around the world.
- **A.** Air pollutants released in Russia can be delivered to the United States in only 3–5 days.
- **B.** Pollutants from the legacy cities around the Great Lakes contribute to acid rain in various parts of the United States.

VI. There are essentially 3 sources of air pollution, all of which are related to energy consumption.
- **A.** Primary pollutants are those that are released directly from the source and that are harmful in that form (e.g., carbon monoxide).
- **B.** Secondary pollutants are those that become dangerous when modified by the environment (e.g., ozone and hydrogen peroxide).

C. Additional sources of pollutants are things like fugitive emissions—particles and chemicals that do not go through a smokestack (e.g., dust from soil erosion, strip mining, and rock crushing).

VII. Even very small particles can have profound health effects.
   A. Very small particles can impact our cellular metabolic machinery.
   B. These fine and ultrafine particulates kill around 500,000 people worldwide per year.
   C. Our government regulates particles for 3 sizes: coarse, fine, and ultrafine.

VIII. There is increasing awareness and emerging consumer demand for clean, sustainable energy options around the world.
   A. Solar, geothermal, wind, tidal, hydroelectric, and even biomass sources are becoming part of the conversation.
   B. So far, the impact has been relatively small, but there is a growing trend.
   C. Industry has responded to the fear that fossil fuel availability will decline by becoming more efficient.
   D. One very interesting alternative is advanced wood combustion, which is a source of renewable fuel.

Suggested Reading:

Alcock, *Animal Behavior*.

Friedman, *The World is Flat*.

Wright, *Environmental Science*, chaps. 12–14.

Questions to Consider:

1. What are the key patterns of human energy use worldwide?
2. How does advanced wood combustion alter our current patterns of global energy use?

Lecture Eight—Transcript
Humans as Energy Consumers

Hello, and welcome back. In the last lecture, we were investigating thermodynamics and its implication for the structure of ecosystems. In this lecture, we'll focus on the ecological reality of thermodynamics and its implications for human survival. Humans are the most import external forcers in ecosystems. We have a voracious demand for energy, and our patterns of energy consumption have long lasting impacts on both local and global ecosystems. Our consumptive patterns produce enormous short and long-term challenges to ecological sustainability.

We're going to investigate energy as the facilitator and limiter of life on Earth. We'll peek into alternative ways of extracting stored energy. We'll focus on advanced wood combustion.

Remember in our last lecture when we looked at the squirrel and considered the hoarding behavior that the squirrel goes through, and I said that some of the nuts that this squirrel buries, he's not going to find them again. In fact, other animals do and, in a sense, that mass crop, both that's there in the ecosystem and hidden by squirrels, is an energy fuel for wintering birds and mammals. They are extracting energy from the system. And so, in a sense, we, like other organisms, will extract energy from the system that's stored.

Now, human demand for energy is increasing, and its acquisition has significant consequences. Consumption of energy and the challenges of delivery are linked to human population growth and redistribution over the past 500 years. Populations have grown and moved to the New World where technology and consumption has skyrocketed. Current world consumption of fossil fuels is absolutely staggering. Somewhere in the vicinity of 85 million barrels of oil are consumed worldwide each day. Daily, in the United States, we consume about 19 million barrels. Each year, 6 billion tons of coal are used worldwide. The U.S. consumes about 1 billion plus tons, and our use of coal is increasing.

If we take a look at data from the Department of Energy, we can see our relationship to energy consumption. You can see in the red line that energy consumption is continuing to grow from an amount in the 1950s of under 40 quadrillion BTUs of energy to the current levels, which exceed over 100. You also note that for quite a significant portion of our time in our expanded technology, our production of energy internally from coal, oil, and other sources, actually matched our consumption. But beginning in the late 1960s and early 1970s, our ability to maintain our own needs has fallen, and our production levels have tapered off, but our consumption has continued to grow. Therefore, our only choice is to import more of the energy that we need. It's a significant change with respect to the pre- and post-World War II era.

Another interesting fact is to think about the energy consumption per person. Here we also look at additional Department of Energy data that go back to 1950. In 1950, just after the Second World War, humans in the United States were consuming about a little over 200 million BTUs each year. That grew steadily as technology increased, and the consumption of typical Americans also increased in response to expanding wealth.

By the beginning of the peak, we peaked at about 1978–79 at about 360 million BTUs, but a series of energy crises and a push within industry to improve efficiency resulted in a bit of a drop in the following decade, followed by a leveling off. Our current consumption is somewhere in the vicinity of about 330–340 million BTUs of energy per person per year, which is a staggering amount, absolutely staggering amount to be able to meet.

This increased energy consumption from fossil fuels burdens the Earth's ecosystem. It increases the amount of greenhouse gas we produce. It increases the habitat loss and fragmentation that ecosystems have to bear. Some of this comes from extraordinary activities like mountain top removal, which is a mining technique that we're using in the United States to extract coal, and also oil extraction itself as the ecosystems that we need to move into to get that oil are more fragile and have more disturbance as a result of the extraction.

Habitat destruction from loss and redistribution of fresh water supplies is also absolutely extraordinary because linked to this demand for energy is a voracious demand for water. We'll investigate water in real detail later in the course, but it's important to consider it here in relationship to energy consumption because often the two are quite tightly linked. For example, the Three Gorges Dam in China. To produce this dam, there was a displacement somewhere in the order of 1.2 million people. But when fully functional, the dam will produce somewhere in the vicinity of 18,000 megawatts of electricity at peak capacity. However, the reservoir behind it stretches back 40 miles and holds somewhere in the vicinity of 5 trillion gallons of water.

The water level in the Gorges will rise somewhere about 100 meters over time and cover somewhere in the vicinity of about 200 square miles. Its impact has spawned landslides and increased risk of earthquakes. This is an example of a balance, the tradeoffs that we talk about in ecologist systems. The dam will, however, offset somewhere in the vicinity of 50 million tons of raw coal that China would otherwise use to produce the electricity. That will have an impact of preventing a certain amount of acid rain from being produced, and it should improve health standards by reducing the amount of air pollution.

It will also become the world's single largest producer of renewable energy. But the overall costs of this system are complex because at the same time we're seeing human displacement, and we're seeing a variety of other implications, not the least of which is changes in the landscape.

Patterns of energy consumption by humans have long-lasting impacts. Of prime concern to us is the increase in greenhouse gases due to our consumption of fossil fuels. Historic dependence on fossil fuels has resulted in a significant burden to the environment. If we take a look at another figure from this Department of Energy report, you can see that the various forms of energy use have grown dramatically over the past century. We've had an increase, then a decrease, and now an increased reliance on coal. We've seen an increase in some renewable sources of energy, such as hydroelectric power, but a tremendous reliance on natural gas, and a tremendous reliance on oil. And so these combined reliances have significant impacts. In fact, our dependence on fossil fuel is ecologically crippling.

The average surface temperatures around the globe have increased somewhere in the vicinity 0.6 degrees Centigrade during the last hundred years. The 1990s were the warmest decade on record, and 1998 was the warmest year in the Northern Hemisphere from data we have that goes back over 1000 years.

Over the last 50 years, the night time minimum temperatures have increased by about 0.2 degrees Centigrade per decade, and there has been a 10% reduction in snow cover ice since late 1960s. The less fresh snow that falls on our ice caps, the more heat absorption occurs in those areas because the snow is actually darker due to dust particles and particulates.

There has been a reduction of about two weeks in the annual duration of lake and river ice over the 20th century. Mountain glaciers have retreated significantly during the 20th century. This has a significant impact on what we call the "albedo effect," and I should take a moment to explain that. It will come up again, but I think it's an important consideration.

Albedo is essentially a measure of the reflectance of a surface with respect to sunlight. In fact, it's the reason that there's a movement in urban areas to paint roofs white because they will reflect more heat, and the building will heat up less. And snow and ice have that capacity as well. But as snow begins to melt and reveals the darker land underneath, then even more heat is trapped, and it acts as a positive reinforcement and literally increases the rate at which the Earth is heating.

In fact, it's the idea why you have road surfaces in temperate areas that are very dark in color mixed into the macadam or pigments that make it a very dark color, almost a black. And so in a snowstorm once you've plowed the roads and you have little bits of pavement showing, that heats up and melts the snow around it, which heats up more and melts more snow, and you get this runaway effect, and so it helps roads clear faster, which is an advantage. But when we're talking about global warming, reducing the albedo effect of surfaces is a significant cost, and it's one of the reasons that the loss of glaciers is so important.

Now, there are, in addition, significant pollution impacts from petroleum, and they can be devastating to marine and terrestrial ecosystems. The impacts of oil spills last for decades. Even well-funded recovery plans like those in Prince William Sound leave lasting negative legacies. If you remember, the tanker Exxon Valdez struck ground in Alaska in March of 1989. It spilled about 11 million gallons of oil that eventually covered somewhere in the vicinity of about 12,000 square miles of ocean. This caused a near total collapse of the local sea populations of clams, and herring, and seals, which also caused two fishing corporations to go bankrupt.

A recent study headed by Joan Bradduck from the Institute of Arctic Biology found that of the oil spilled—this is very interesting—14% was recovered, 13% is still in the sediments and beaches. About 1% remains in the water, about 20% evaporated, but 50% underwent biodegradation, which means microorganisms literally consumed the oil as a food resource and released it in less toxic forms. This is another example of the extraordinary resiliency that ecosystems have, even under the most extraordinary conditions. Remember, in a previous lecture we talked about recovery of anoxic dead zones in ocean ecosystems. This is another example.

Reliance on petroleum has increased global political tensions. Over the past 100 years, there have been significant resources expended, and loss of life, due to protection of our nation's access to oil resources. This cycle intensifies as the resource becomes scarcer.

According to Royal Dutch Shell's Governing Committee Chairman, Jeroen van der Meer, the relationships between governments will be the critical factor in the two-part challenge of delivering critical fossil fuel reserves to hungry nations; that, and the confrontation of climate change. He considers those to be of equal challenge. The geopolitics of oil are in the thick of the ecology of the world's future.

Now, extraction technologies and pipes must enter and cross many national borders. As a result, small nations can exert tremendous influence on the world by hindering or disrupting petroleum deliveries. This system is likely, or this conflict is likely to increase, as we move into even less stable political environments and more fragile ecosystems to extract remaining fossil reserves.

Let's think a little bit about oil and this ecological legacy of pollutants. Remember, we talked about Prince William Sound. It's just one example of what we call the "legacy of pollution."

Pollutants released into the atmosphere can travel around the world. This transport efficiency of pollutants is calculated as transit time. So pollutants released into Russia as air pollutants can be delivered to the United States in only 3–5 days. I know that sounds extraordinary because Russia is halfway around the globe, but let's take a look at how this actually takes place.

Here you see an image of the globe, and superimposed on that image are the dominating air currents. Remember, the air currents are the result of a variety of large-scale interactions that happen on the globe. The heating and cooling of the globe is a result of the impact of solar energy, the Coriolis effect generated by the spin of the globe, and so forth, but you can see by looking at these dominating air currents that there is an almost immediate delivery from these centralized industrial areas to other parts of the globe. And it doesn't take long for pollutants to move from one area to the next.

Pollutants from the legacy cities around the Great Lakes, that we pejoratively called the "Rust Belt," contribute to acid rain around various parts of the United States, including the Appalachian Mountains. We call them "legacy cities" as they are historically old, and they shaped the industrial ecosystem of the United States. If we consider this legacy from the standpoint of fates and movements of chemicals in the environment, we get an entirely different, but enriched picture.

There are essentially three sources for air pollution, all of which are related to energy consumption. One source of air pollution is what we call "primary pollutants." These are released directly from the source and are harmful in that form. Typically, they're related to combustion technologies, things like carbon monoxide, which is poisonous at the time that it's released.

There are other kinds of products that are released that we call "secondary pollutants." Those become dangerous when they're modified by the environment and made hazardous after they have been released. Those are such chemicals as the photochemical oxidants and atmospheric acids. Photochemical oxidants include ozone and hydrogen peroxide, and as they engage additional things like atmospheric acids and sulfuric acid, they become toxic as they fall back to Earth.

Additional sources of pollutants are things like fugitive emissions. These are particles and chemicals that don't go through a smokestack. For example, dust from soil erosion, and strip mining, and rock crushing, those particulates provide an additional source of pollutants. In the U.S. it's not insignificant. This results in 100 million metric tons per year.

Worldwide emissions of carbon based pollutants went from about 1800 million tons in 1950 to over 7000 million tons at the turn of the century. These data cause particular concern in light of recent research that even very small particles can have very profound health effects. These particulates kill from lung disease and mitochondrial diseases that are related to these very small particle sizes.

A.E. Nell at the University of California at Los Angeles recently published in *Science* that very small particles can end up impacting our cellular metabolic machinery. Worldwide these fine and ultra-fine particulates kill somewhere around 500,000 people per year. This is an extraordinary figure.

Our government regulates particles for three sizes: coarse, fine, and ultra-fine. Ultra-fine particles are less than 0.01 micrometers in diameter and are released as a product of combustion, typically diesel fuel. So these microparticles are a whole new ecologist dimension of challenges that we need to investigate.

Now, what about alternatives? It's clear that our economy is based on our ability to provide energy for necessary services, and there is an increasing awareness and emerging consumer demand for clean, sustainable energy options around the world. Solar, geothermal wind, tidal, hydroelectric, and even biomass sources are becoming part of the conversation. Not just in laboratories, but also among concerned members of the general public.

So far the impact has been relatively small, and if we take a look at additional data from the Department of Energy, you can see that the demand in use for fossil fuels has grown voraciously since the 1950s, with a slight drop in response to the concern over the availability of oil that happened in the 1970s, but it's still continuing to grow. If we look at renewable sources and also nuclear electric power, we see that there is a growing trend, but it's still a relatively small proportion. So the question is, where are these alternative sources going to come from?

Now, the ideas are driven both by forward thinking about sustainability, but also this response to the fear that fossil fuel availability will decline. Industry has responded to the stress by becoming more efficient. If we take a look at an additional figure from this report, we can see that the energy consumption per real dollar of gross domestic product has actually reduced over time. The companies that are responsible, or the industries that are responsible for productivity, have recognized the increased cost and are doing their best to try and save upfront production costs. That has resulted in becoming more efficient. And so it's interesting to see that if industry can do this, at the consumer front we ought to be able to do so as well.

One very interesting alternative is something called "advanced wood combustion," a source of renewable or biomass fuel. There was a time in the history of the United States when wood provided more energy than fossil fuels. However, we tend to consider wood an old and dirty technology. But it turns out not to be so, according to a team led by Daniel Richter at Duke University. He argues in the *Journal of Science* that they propose a global reconsideration of this emergent technology that utilizes wood in a renewable fashion to generate both heat and electricity.

Building on data from Europe, especially in Scandinavia, they report that these systems are generally small-scale and decentralized. Some run as high as 90% efficiency, and some countries, one of which is Austria, have embraced these power plants and built over 1000 of them, 100 of which are actually producing both electricity and heat.

So an emerging question is, can these systems be adopted for use in the United States in a sustainable fashion? Now, Richter admits that this is an open question with hurdles that are more social than technical.

According to the Department of Energy, we consume about 100 quads of energy each year. A quad is 100×10^{15} BTUs, or 1 million times 1 million BTUs. Wood currently provides somewhere in the vicinity of two quads, but that could easily, according to these researchers, be expanded to five quads. Now, to put this into perspective, hydroelectric production is somewhere in the vicinity of three quads and our entire strategic petroleum reserve is about four quads. So the amount of energy that could be produced by these advanced wood combustion technologies is significant.

Now, with Americans generating over 350 million tons of wood per year that gets used primarily for energy generation, advanced wood combustion can provide an additional sustainable harvest of wood products. The trick here is the balance within each ecosystem in which the extraction activities are undertaken.

The controversial piece is the release of carbon when the trees are burned, which does inevitably add to greenhouse gas emissions. However—and here's the interesting ecologist part—the team argues that these carbon sources are already in the ecosystem, and that they are not coming from deeply sequestered sources such as oil or coal, and that the combusted carbon will be re-sequestered by the growing forests that will supply the next round of these advanced wood combustion technologies.

The economic equation is more favorable, in fact, in the United States than it is in Europe. They favor essentially three key steps in making this work. One is to make advanced wood combustion the preferred choice for new construction in areas that are favorable. In other words, if people are applying for permits for new homes and buildings and are already using the kind of clustered technology and building activities and smart growth demand, something we'll look at at the end of this course, then using these advanced wood combustion technologies would make sense. And if they're required at the time of construction, then you don't have the cost of conversion, which is a significant calculation that happens when you're asking people to change energy forms. In fact, the cost of AWC, as it's called, is buried into the general cost of developing the system to begin with.

The second part of the strategy is to be more efficient with waste wood. We generate enormous amounts of waste wood in this country each year, and if there was a system in which it could be utilized more efficiently, then there would be the impetus to save that wood and use it for these AWC technologies.

The final sort of strategic piece here is to develop more district level energy systems where heat is from a central source, which tends to be something that universities, and hospitals, and other types of organizations have done historically, but as we went to these big mega-energy systems in the 1950s, '60s, and '70s, we moved away from that direction. However, certain cities like Akron, Ohio, and St. Paul, Minnesota, and some universities such as Colgate University and the University of Idaho, have adopted these AWC strategies.

Now, when we think about the challenge of energy consumption, I can't help but be reminded in my own native Massachusetts to think about the decline of the whaling industry. At one time whale oil was prized for the enormous amount of heat energy available within the oil. It was also an extraordinary pre-petroleum product with respect to its lubrication capacity, and so forth. And the acquisition of whales and the resources that they provided made Massachusetts and some of the communities where whaling was intensive, like New

Bedford and Nantucket, were some of the wealthiest communities in America because of this availability of the resource.

But it didn't take us that long, really, only about 100 years or so, to actually drive that resource nearly into extinction. We tend to think of living organisms as being renewable, but we actually managed to fish whales almost to extinction. As such, the resource became prohibitively expensive, and the emergence of new technologies, which was the discovery of oil in Oklahoma, Texas, and Pennsylvania, suddenly, in a very short amount of time, made whale oil essentially obsolete. We saw the demise of that industry in Massachusetts, which was not really prepared in many of those cities to move to the next phase of industrial development.

One final piece I would bring forward in our conversation is that we've actually learned quite a bit about the role of microorganisms in maintaining healthy ecosystems and the development of oil-eating bacteria, which allow us to use bacterial infusions into areas where oil spills have taken place to help clean up the system. We learned a bit about that with respect to Prince William Sound. But here's the interesting piece. It also turns out that you can use these same kinds of oil-eating bacteria, inject them into relatively low quality ecosystems that are storing oil, like in shale and oil tar sands. In these oil tar sands, which are very hard to extract the oil from, it turns out that if you inject these oil-eating bacteria they can transform the oil into forms that can be extracted with less ecological damage.

Now, when we next gather, we're going to look at the material side of the equation when we consider the recycling of nutrients in our ecosystem. So, until then, farewell.

Lecture Nine
Nutrient Cycling in Ecosystems

Scope:

Unlike energy, which passes through our ecosystem and leaves as heat, the materials of life are reused over and over. In this lecture, we examine the movement of materials through ecosystems and how the recycling of matter can enhance an ecosystem's physical resiliency.

Outline

I. The materials of life get used over and over again.
   A. The supply of these organic molecules is finite.
   B. The availability of nutrients actually limits the size and distribution of organisms.
   C. We literally are what we eat: When a coyote eats a bird egg, the carbon locked up in the yoke of that egg will be broken down and rebuilt and will wind up in the tissues of that coyote.

II. The role of humans as hunters is very important in ecosystems.
   A. In New England, before there were cities, the dominant top-order predators were bears, mountain lions, and wolves.
   B. As humans moved into New England, they displaced those other top-order predators and altered the energy flow.
   C. As fewer and fewer humans in urbanized areas hunt, we see an increase in the deer population because of a release in the top-down pressure; this has resulted in increases in zoonotic disease and the number of deer being hit by cars.
   D. These are social and behavioral decisions that humans are making, but they have profound impacts on nutrient cycling and energy flow.

III. All materials in ecosystems are recycled and reside in compartments.
   A. The world has a finite amount of any given nutrient, and the speed at which it moves through ecosystems is highly variable.

- B. Cells take chemical energy and turn it into physical tissue. This acquisition of chemical materials and their transformation into more complex forms is also at the core of the evolutionary process.
- C. The individual organism is just a fleeting aggregation of organic molecules.
- D. The genes in our bodies are much older than we are, and the materials in our bodies will be recycled after our passing.

IV. Now we return to our energy diagram from the previous lecture.
- A. We see a cycle between the physical environment and the organisms that live there.
- B. Humans acquire carbon and other materials by eating producers directly or eating some animal that ate a producer.
- C. We will, at the time of our death, and also by eliminating waste, release unused materials back into the ecosystem.
- D. The process of nutrient cycling is a constant exchange between the physical environment and living organisms.

V. Geophysical processes such as erosion and sunlight physically alter materials in the ecosystem.
- A. These are abiotic forces releasing nutrients into the ecosystem.
- B. When I was studying goats in British Columbia, I was amazed at the distances they would travel to acquire precious minerals that they used as metabolites.
- C. They were accessing the salts that were released from the weathering rock and from the forces of wind and rain.

VI. Human technological practices have dramatically sped up the process of materials cycling.
- A. Mining and combustion processes accelerate the release of materials into the biosphere.
- B. High-tech activities produce highly toxic materials.
- C. One obvious example of this is e-waste, which is the junk created when our electronics and computers wear out.

VII. The key molecule in life is carbon it its many forms.
  A. Carbon is held in 4 reservoirs: the atmosphere, the oceans, sedimentary deposits, and dead organic material.
  B. Looking at the carbon cycle, you can see the movement of materials from the soil into the atmosphere, into the sediments, into the ocean, and so forth.
  C. When areas are deforested, a profound amount of carbon is released back into the air.

VIII. One of the biggest impacts of humans with respect to geologic forces occurs because we intervene in the carbon cycle.
  A. The petrochemicals buried deep beneath the surface of the Earth would normally be there for millions of years, but we dig them up for energy.
  B. The amount of carbon being released into the atmosphere has increased dramatically over the past 200 years.
  C. Rising amounts of carbon dioxide in the atmosphere are causing a rise in surface and sea-level temperatures.

Suggested Reading:

Cain, Bowman, and Hacker, *Ecology*, chap. 21.

Scherr and McNeely, *Ecoagriculture*.

Questions to Consider:

1. How does the movement of materials through ecosystems compare to the movement of energy?
2. How do materials such as nitrogen move through ecosystems and transition from nutrient to pollutant?

Lecture Nine—Transcript
Nutrient Cycling in Ecosystems

Hello, and welcome back to our investigation of ecology. In the past few lectures, we have talked about the movement of energy. We've talked about the ecological side of this, the challenges, the pollution impacts, and even some of the geopolitical issues related to acquiring energy as humans.

The other side of the equation, of course, is the materials. In previous lectures, we set the stage by saying that energy flows in a unidirectional way through ecosystems, arriving from the Sun, transformed into a variety of chemical holding patterns, if you will, used to run the metabolism of organisms: reproduction, repair, movement, and so forth. Some of it gets passed on to the next trophic level, but ultimately, that energy is lost to the ecosystem as heat.

But organisms are more than just their energy. Organisms are made up of stuff. And so where does the stuff come from? In this particular lecture, we're going to be focusing on the idea that the materials of life get used over and over again. These organic molecules are finite. There is not an endless supply worldwide. The availability of nutrients actually limits the size and distributions of organisms. There tend to be particular limiting factors.

We're also going to consider how the recycling of matter can enhance an ecosystem's physical resiliency. As the old adage goes, "You are what you eat."

We're going to begin our conversation today with a couple of images, both of which come from our study site down on Cape Cod. In the upper left is an image of an endangered shore bird called the "piping plover." It was actually the focus of my doctoral studies back in the 1980s, and it was a community-based study in which we organized students and local participants, as well as scientists, to investigate a long-term study of the impact of humans on piping plovers.

It turns out that piping plovers and humans want to use the same barrier beach corridors. The plovers need them for nesting, and humans need them for recreation, which is a very important economic engine for coastal communities like those on Cape Cod.

What I was really doing was studying ecologist conflict of competition among species. But in the previous lecture, remember, we said that there can be direct and indirect forces that one species exerts on the other. Here is where I think this picture is particularly interesting. Piping plovers nest on the shore right near the ocean edge, above the high tide line, and they lay their eggs right in the sand. Their eggs are camouflaged so they're very hard to see, and that works pretty well. We're going to actually investigate later in the lecture series some of the details of this activity that they go through for their reproductive life cycle.

The picture on the bottom is the one that's so fascinating to me. This is a remote camera that's set up in front of a piping plover's nest. Historically, in any given year, about one-third of the nests were lost to predators. They didn't even hatch. That's not unusual for species that live along the coastal margin. It's living in an area that's very exposed, and predators begin to key in on the areas in which these prey species are found.

The irony here, of course, is that our lab also studies coyotes, which we mentioned before, and here is a game camera picture of a coyote taken at night. There's a flash that's happening here so we can tell what animal it is. This is a coyote discovering a piping plover nest and about to devour it. Why this is relevant to this conversation today is that this is an exchange of energy. But it's not just an exchange of energy. It's an exchange of materials. So, when the coyote consumes the eggs that were laid by the piping plover, not only will it break down the chemical materials in those eggs, the complex carbon compounds, and extract energy that was locked up in the bonds of those chemicals, but it will also extract materials that it will use to repair and build its own body. So, literally, the carbon that was locked up in the yoke of that egg will be broken down and rebuilt, and will find itself in the tissues of that coyote. We literally are what we eat. It's not a metaphor.

This links also to an additional consideration before we jump in, in that this is a measure of trophic dynamics. This is a bridge from what we've done in the past couple of lectures to what we're going to do in the next couple of lectures. These animals in the picture here are at different trophic levels. But the picture that's missing here at the very top of the trophic level in this system is humans. Humans not only live with coyotes, but occasionally hunt coyotes, and so it's another level within this trophic system.

In fact, the role of humans as hunters is actually a very important one in ecosystems, especially in those ecosystems where humans are the dominant species. If we think historically about a place like New England, where this research is taking place, if we go back 250 years before there were cities, then the dominant top order predators in a community like that would have been bears, mountain lions, and wolves. But as humans moved into New England in huge numbers from Europe, they displaced those other top order predators and, in essence, this interaction, this energy flow, this movement of energy from one trophic level to the other, was altered.

Just like we learned in the previous lecture, the direct and indirect effect of predators is incredibly important. When you lift those predators off, it changes the ecosystem, and as strange as it may seem, the remaining top-down effects that those predators would have exerted on an ecosystem—the wolves, the bears, and mountain lions—are now being exerted by a relatively few number of species. Coyotes are one, but humans actually become a very important top-down force.

For animals like deer, which are relatively recent inhabitants of North America, only about maybe 20,000–30,000 years ago, humans have always been, and continue to remain, probably the most important predator of deer. As fewer and fewer humans in urbanized areas are hunting, we're seeing an increase in deer population because of a release in the top-down pressure similar to the way that the pressure was released when moose populations dropped in Isle Royale. We're seeing that the balsam firs were released and were growing faster. As hunting pressure and predatory pressure on deer are released, then the deer population grows. What are the results? We're seeing increases in zoonotic disease, increases in the number of times that humans in their cars are hitting deer, which now is over a quarter million times each year in the United States.

We're seeing a system that is changing as a result of changing demographics and consumer patterns of humans. These are social and behavioral decisions that humans are making, but they have profound impacts on the nutrient cycling and the energy flow. We are on the bridge. We are now crossing from a conversation of energy flow into a conversation of nutrient cycling, and, boy, are we seeing it here. Those nutrients are moving from the piping plover into the coyote.

All materials in ecosystems are recycled and they reside in compartments. Compartments are sort of metaphorical, but physical, domains so that we can sort of consider how these chemicals move. The world has a finite amount of any given nutrient, and the speed at which it moves through ecosystems is pretty highly variable.

Anabolic biological activity increases the complexity of chemical substances. Using chemical energy, cellular metabolism can take basic materials to produce living structures. This is the essence of how metabolism works. Cells take chemical energy and turn it into physical tissue. This acquisition of chemical materials and the transformation to higher complexity is also at the core of the evolutionary process. However, life spans of all organisms are also finite, and organisms do not live forever.

Individual organisms may pass on their genes. This is how ecologists define the notion of fitness. Remember, going back to our Darwinian conversation a couple of lectures ago, natural selection favors populations of organisms, favors individuals within those populations because of the traits that they're expressing at any one time, and how those traits mesh with the dynamics of an ecosystem. Those individuals that reproduce more frequently have their genetic material represented in the population in the next generation to a higher degree.

However—and I don't mean to sound philosophical because I'm trying to be concrete here—the individual organism is what we call a "fleeting aggregation" of organic molecules. More cynically, we are a bag of DNA that exists for a certain amount of time, and we either pass on that material to the next generation or we don't. But our own lives in the organism that we inhabit now comes to an end, and it is a reality that is difficult for us to consider because our cognition is around our own lifetime. But the genes in our bodies are much older than ourselves, and the history of humanity goes way back, much before the beginning of our own life span, and continues on beyond our own life span. The materials in our bodies will be recycled, either quickly or slowly depending upon the decisions we make at the time of our own passing.

When organisms die or eliminate waste, these materials are broken down by organisms, and that group of organisms we call "decomposers." We introduced them in the previous lecture. The ecosystem services they provide involve the releasing of those complex molecules back into the physical environment where they can be utilized by other organisms in the future.

What I want to do now is return to our energy diagram that we began to investigate in the previous lecture. Remember, on the left side we showed the movement of energy. Beginning with the Sun, solar energy is absorbed by the group we call "producers." That energy is captured into the chemical bonds of particularly light-sensitive pigments. That energy is now trapped into biological tissue, and is now available to other organisms. And so we see how energy trapped by producers ends up fueling the metabolism of consumers. Ultimately, that energy passes through and is lost as heat.

Notice on the right side of the diagram. The left side of the diagram has a straight line from the Sun to heat. But on the right side of the diagram what we see is a cycle between the physical environment and the organisms that live there. That cycling involves the movement of matter because the carbon and other materials in our bodies were bio-accumulated by other organisms in previous lifetimes. We acquired them by eating producers directly or eating some animal that ate a producer. We either ate an animal that ate plants, or we ate plants directly, and we incorporate those carbon skeletons into our own framework. And we will, at the time of our death, and also by eliminating waste, release those unused materials back into the ecosystem. There is a constant exchange between the physical environment and living organisms. The energy is flowing in one direction, but the cycle of materials is continuous.

That's important to remember because ultimately the amount of complex molecules that is available is a very, very small fraction of the matter in the universe. Most of the matter in the universe is hydrogen and helium. You can't build much of an organism out of those inorganic components. But the carbon-based molecules that are out there are actually very, very rare, and it's living systems that bio-accumulate them. But we don't get to keep them forever, and we pass them on through this process of nutrient cycling.

The geophysical processes such as erosion and sunlight physically alter materials in the ecosystem. This is an example of abiotic forces releasing nutrients into the ecosystem. When I was in graduate school, I was fortunate enough to accompany my research colleagues to a site in northern Canada where they were doing a long-term study of mountain goats. It was being conducted by a Tufts University ecologist by the name of Benjamin Dane. It was awesome country with towering peaks that, at first, really overwhelmed me. I'm a coastal flatlander, and our base camp was at 11,000 feet. It actually took me quite a few days to overcome a sort of uncertainty about heights, and to get used to working in what is an extraordinary rugged and beautiful countryside. In fact, for the site we worked at, we went in by float plane and then by helicopter, and in fact, from the tops of these mountains it was such a remote location you couldn't even see roads from the tops of the mountains, let alone buildings. It was a transformative experience for me, growing up in heavily populated areas of the Great Lakes region, and then on the East Coast, and then Europe for a while. I really had never been in an environment where I hadn't seen houses and people constantly.

What amazed me the most when I was studying these goats in British Columbia was that they would travel great distances to acquire precious minerals that served for them as metabolites. You literally can think of them as goat vitamins. They were accessing the salts that were released from the weathering rock and from the forces of wind and rain. It's rugged, rugged countryside up there. In fact, in future lectures we're going to come to understand how these different kinds of biomes, especially those in mountains, get formed.

Back to the goats. This is really risky business for the goats because when they come down off the mountain slopes into the valleys in order to lick these stones, there's a very, very high risk that they will be preyed upon by wolves because down in the flat valleys the wolves can move very quickly, and the goats are exposed. But it's an interesting payoff because the nutrients are essential. So, the goats have this challenge. They can stunt their own growth and reduce their probability of reproduction by staying on the mountain, or they can get the salts that they need by coming down and taking a big risk. So, their need for salt overcame their fear of wolves. We'll actually look more closely at the role of wolves and change in the behavior of prey species later on. It's a fascinating story coming out of Yellowstone, which we actually had touched on before, this same kind of ecology of fear.

From the remote areas of British Columbia back to human technological practices, which have been a focus for us as we think about the sort of crisis aspect of our work in ecology. Human technological practices have altered the temporal aspects of materials cycling. We've speeded it up dramatically. The compartments that these nutrients would likely sit in, or materials, we've increased the rate at which these materials move.

Mining and combustion processes accelerate the release of materials into the biosphere. High tech activities, as we mentioned before, produce highly toxic materials. One of the fascinating aspects of technology is that we can create heat and pressure that can concentrate highly toxic materials into small spaces. Then, unfortunately, we often release them into the biosphere. One obvious example of this is something we call "E-waste," which is the junk that happens when our electronics and computers wear out, which given the voracious consumption of electronic devices that we have in developed nations, we turn over this material incredibly fast. We're going to actually talk a lot in the next lecture about the whole issue of E-waste and human garbage. Suffice it say for now that our appetite for electronic technology has resulted in mountains of toxic trash. Unfortunately, that burden falls disproportionately on poor nations who attempt to make a living trying to recycle this material.

The key molecule in life is in no doubt—it's carbon in its many molecular forms. The carbon cycle is very important for the movement of carbon, which is a vital nutrient, from the Earth, in and out of the atmosphere in various compartments, and back to the ground. We're going to look at the carbon cycle in some detail.

When we think about carbon, we can consider it either as it is exposed to human access, or it is sequestered. If it's sequestered, it is essentially not exposed to humans and not accessible. Carbon is essentially held in four reservoirs. The first is the atmosphere, primarily as carbon dioxide. And remember, we've already had a conversation and will re-engage this about the role that carbon dioxide, as a gas, plays in the overall climate conditions of the planet.

Plants take about a quarter of the carbon dioxide out of the atmosphere, and they photosynthesize it into carbohydrates. Remember, that's the key step in making life available on this planet. Making life possible is the process of carbon capture that happens in photosynthesis as the combination of sunlight, water, and carbon dioxide get transformed into carbohydrate.

About another quarter of the world's carbon is absorbed by the world's oceans in carbonate rocks, in the soils, and also in the sediments. These are compartments that can last for a significant period of time. Limestone and coral are examples of critical warehouses of carbon, so limestone and coral, especially coral, are generated through biologic processes. And locking up carbon dioxide in coral has some very interesting implications not only for climate, but also for buffering the changes in acidity in marine ecosystems.

Carbon is also found in sedimentary deposits, such as coal, petroleum, and natural gas. Remember, we talked about the challenges in the last lecture of trying to extract these materials from the Earth and using them as energy sources.

Finally, carbon is stored as dead organic material such as hummus in the soil. So, if we take a look at the carbon cycle, you can see a series of arrows that help to illustrate this movement of materials from the soil into the atmosphere, into the sediments, into the ocean, and so forth. We can see exchange happening directly from the ocean to the air as carbon dioxide is exchanged among the coral communities in the atmosphere. We can also see carbon dioxide entering the atmosphere as a byproduct of combustion, as takes place in both automobiles and in factories at any time that hydrocarbons are put under high pressure and heat and break apart.

We can also see that there are natural sources of carbon dioxide that come from combustion, such as volcanoes. When volcanoes take place they release enormous amounts of carbon dioxide into the atmosphere.

Don't forget, you can also see the arrows here that connect the atmosphere to plants. Plants exchange carbon dioxide in the atmosphere. They take carbon dioxide in as part of photosynthesis, but remember—and this is something that many people forget, including my students and sometimes even myself—we tend to get so focused on the fact that plants take sunlight and do photosynthesis, we forget that plants also have to survive at night, and they have to survive in winter when they drop their leaves. So, the process by which they acquire these carbon reserves and build them into complex molecules, they're not doing that to serve us as consumers. They're doing that to serve their own metabolic demands. And so, not only do plants photosynthesize, but plants also go through energy production just like we do and give off carbon dioxide as a waste gas. It's easy for us to forget that sometimes plants are absorbing carbon dioxide and sometimes they're giving it off.

When areas are deforested there is a profound amount of carbon that is released back into the air through this process because not only will some of the trees be burned by humans to clear the land. There will also be the breakdown of all of that cellulose, the wood that breaks down, which are really just long carbon compounds, and those will be released into the air as carbon dioxide. And don't forget, humans and other animals release carbon dioxide into the atmosphere by respiration and breathing, especially as professors, you know. We're noted for all the hot air that we produce.

Carbon is released through the process of weathering of rocks, as well as the decomposition of plants and animals. We can think of that carbon cycle as divided into two categories: the geological part, which happens over millions of years, weathering and erosion; and the biological part that works over a shorter period of days and centuries.

One of the biggest impacts of humans with respect to the geological forces occurs because we sort of intervened in that process. All of those petrochemicals that are buried deep beneath the surface of the Earth would normally be there for millions of years. But we dig them up, burn them to produce either heat, or electricity, or motion, and that releases the carbon back into the atmosphere more rapidly than it would normally happen.

The amount of carbon being released into the atmosphere has increased dramatically over the past 200 years. Almost all of us now are familiar with the so-called graph of atmospheric carbon dioxide concentration that was made famous by former Senator Al Gore in his film, *An Inconvenient Truth*. We're devoting an entire lecture to greenhouse gases, so I'm not going to spend much time on it now. But the overwhelming majority of climate scientists believe this increase is mostly due to anthropogenic use of fossil fuels. Although, remember, agriculture and cattle come in a close second with respect to the impact and release of carbon dioxide.

Rising amounts of carbon dioxide in the atmosphere are causing a rise in surface and sea level temperatures. We're going to devote an entire lecture to this critical aspect of our course.

The nitrogen cycle is also important because the movement of nitrogen is critical with respect to the development of proteins. Along with carbon, it's the key component of proteins, and proteins make up most of what we are as an organism. Nitrogen is the sixth most abundant element on Earth's surface. It makes up about 78% of the atmosphere. Nitrogen mostly exists in its inert form in the atmosphere, and does not cycle throughout the atmosphere and the Earth without the help of living organisms, the so-called "nitrogen-fixing bacteria." These highly specialized bacteria establish a relationship, a symbiotic mutualism, with many plants with respect to their roots in the soil. Nitrogen-fixing bacteria take inert forms of nitrogen and make it available to plants. Typically, these bacteria are found either in the root nodules of some of the plants, or they're associated with the root tissue.

The key thing that they do, again, is take this inert form of nitrogen, which is very common in the atmosphere, but inaccessible to us biologically, and turn it into a more chemically active form. Often, this is the first step in actually colonizing inorganic soils, and certain plants and their symbiotic nitrogen-fixing bacteria are particularly good at that. One species is lupines. Lupines and their symbiotic nitrogen-fixing bacteria that re-colonized Mt. Saint Helens after the volcano erupted in 1980, served to provide this critical metabolic bridge for nitrogen into the soils so that plants and insects could follow.

Nitrification is another process by which bacteria convert ammonia into ions that are more usable by most plants, such as nitrites. This occurs actually only under aerobic conditions and is, thus, limited to terrestrial ecosystems.

Then there are some microorganisms that, under extreme conditions of low oxygen tension, something that we call "anoxia," will actually denitrify the soil. These bacteria use the nitrate ion directly as a key metabolite in their quest for energy. They convert it to either nitrous oxide or back to diatomic nitrogen, and so it's sort of lost to the atmosphere. So, those actually are processes where the nitrogen goes back out of the bioavailability region and into the abiotic world.

Nitrogen can be found in the oceans, atmosphere, in the soil and plants. These nitrogen-fixing bacteria help to move the nitrogen from the soil into the plants by making the nitrogen more metabolically active. Some nitrogen is actually released into the atmosphere through lightning. And nitrogen is released into the air by combustion from factories.

Pollution has tremendous impacts on biogeochemical cycling. Humans have a large impact on these two cycles by moving carbon and nitrogen artificially from one compartment to the next. For example, fertilizers used on crops. That doubles the amount of nitrogen transfer that would be taking place in an ecosystem. The increase of nitrogen oxide especially affects the ozone layer, and has contributed to the depletion of that ozone layer.

The use of fertilizers on crops has tripled the amount of ammonia which contributes both to poor air quality and acid rain. The increase of fertilizer use also causes problems when there is a large amount of fertilizer that runs off into oceans, and lakes, and rivers. It causes eutrophication, and that causes an increase in the photosynthetic activity at first by plants, and bacteria causes an algae bloom that depletes the amount of available oxygen for fish and other consumers. We talked about that before when we talked about the dead zones.

There was a famous experiment looking at the impact of limiting nutrients in the 1970s at an experimental site in Manitoba. A team led by ecologist, David Schindler, from the University of Alberta, investigated the role of limiting nutrients on aquatic ecosystems. One experiment added phosphorus, another critical nutrient, to a series of experimental lakes, and the result was eutrophication from a bloom of cyanobacteria. Phosphorus turns out to be a key limiting chemical element in DNA. These studies helped us to illuminate the importance of nutrients in causing pollution.

Under the wrong conditions, even healthy nutrients can be pollutants and degrade ecosystems. This discovery led to the control and development of new regulations regarding the discharge of phosphates into oceans, lakes, and streams. Nitrogen runoff from fertilizers and animal waste remains a huge problem.

In the next lecture, we're going to tackle the huge problem of human waste and the challenge of recycling, including E-waste. So until then, farewell.

Lecture Ten
The Challenges of Waste and Disposal

Scope:

In this lecture, we examine the nature and extent of human trash, along with its impact on the environment. We discuss the ecology of the end stage of human consumption and consider alternatives to current strategies of waste disposal.

Outline

I. The ecology of human trash is extraordinary for both its complexity and enormity.
   A. Each day in the United States, 90,000 truckloads of trash are picked up and disposed of.
   B. Unfortunately, even recycling efforts often result in ecological toxicity and human suffering.

II. Human waste streams are increasing in volume and toxicity, and their disposition represents an enormous challenge.
   A. In rural areas of the world, human metabolic waste is just buried or dispersed. But in urbanized areas, the amount is so large that we must collect and treat it.
   B. Human waste is treated in waste treatment plants with a combination of chemical and physical processes.
   C. There are traditionally 3 stages of wastewater treatment, which treat both wastewater and street runoff.
   D. When there are large amounts of rain, many sewage treatment plants become overwhelmed, and the overflow is allowed to bypass the treatment process, leading to pollution.

III. Increasing amounts of solid waste, or garbage, is another problem.
   A. There are about 240 million tons of garbage generated annually in the United States.
   B. The percentage of disposability has increased over time as we shift to a use-once, throw-away culture.
   C. Over the past 40 years, the amount of municipal solid waste has doubled in the United States.

- D. The ecological impacts of trash are absolutely devastating.
- E. Historically, cities either buried their trash, burned it, or took it out to sea and dumped it.
- F. In landfills, layers of trash are covered with soil so that microorganisms and bacteria in the soil will aid in decomposition, but soil and water contamination often result.

IV. We have a whole new set of problems related to the disposal of e-waste like computers and televisions.
- A. Inside these machines lies a toxic soup of heavy metals and other poisons.
- B. Those materials are mostly harmless while the equipment is intact, but when we take them apart to recycle them, the toxic materials are exposed.
- C. Even when we do not recycle these objects, they get crushed under the weight of materials on top of them and begin to release the toxic materials.
- D. Americans discard approximately 2 million tons of these electronic devices each year.

V. Marine dumping is another practice used for waste disposal.
- A. It was banned worldwide after debris began washing up on public beaches.
- B. It is very hard to catch the illegal disposal of substances like plastic and medical waste.
- C. In the ocean, trash does not stay where we dump it. It gets moved around by the massive current system.
- D. Trash collects in areas where the currents are weakest and results in huge submerged islands of trash.
- E. Tires have historically been used for coral reef restoration, but over time the rubber deteriorates and releases toxic chemicals.

VI. E-waste and toxic metals are becoming an increasing problem, especially in poor nations.
- A. Most e-waste is not biodegradable, and it tends to be sent to poor countries to be taken apart, due to lower environmental standards there.

- **B.** In India and other countries, people collect parts that contain precious metals and melt them down—often in the same pots they use to cook their meals.
- **C.** A 2005 study found that when African landfills got too high, villages would just set them aflame.
- **D.** A long-term study being conducted in China is finding that children living near e-waste dumps have significantly higher blood levels of lead than those living farther away.

VII. Recycling is still a critically important means of reducing waste.
- **A.** Many people argue that the cost of recycling and removal is actually greater than the use-and-discard strategies that have been implemented over the past 50 years.
- **B.** But recycling in urban areas is more cost effective than use-and-discard.

Suggested Reading:

McDonough and Braungart, *Cradle to Cradle*.

Wackernagel and Rees, *Our Ecological Footprint*.

Questions to Consider:

1. How would you characterize the nature and production of human waste as the species moved into the industrial age?
2. Considering that humans are primarily a terrestrial species, why are the oceans a particular focus of the ecological study of waste?

Lecture Ten—Transcript
The Challenges of Waste and Disposal

Hello, and welcome back. From our previous lecture, we learned that the stuff of life, the materials, is used over and over again. This is also true for human materials. They simply don't disappear, but unfortunately, often we tend to treat them that way, as though they do disappear. The ecology of human trash is extraordinary for both its complexity and enormity. Each day in the United States, 90,000 truckloads of trash are picked up and disposed of. In this lecture, we're going to examine the nature and extent of human trash, along with its impact on the environment.

Even recycling efforts often result in ecological toxicity and levels of human suffering. This has become one of the most vexing challenges of the 21st century. We'll examine the ecology of the end stage of human consumption and consider alternatives to the current strategies of waste disposal.

Think about this. Our patterns of consumption are so locked in, especially in Western culture, that we don't realize how much material we generate. I like to use the example of the overflowing shower stall. I don't know if your family is anything like mine, but you make your way into the shower and there are different shampoos, and soaps, and all these products. And each individual who's using that has their own personal favorites. Not only is this expensive and sort of messy to deal with, but think about it. The chemicals in those bottles, after you've used them on your body and get washed off, go down the drain. The materials that the chemicals came in, the soaps, have to be thrown away, and so that becomes plastic waste. And there was additional energy required to move that material around before it got to you. We have all these packaging and products that are stored for relatively short periods of time between the shelf and our consumption. And then, ultimately, we have to dispose of not only the chemicals, but ultimately, the containers that they came in.

Human waste streams are increasing in volume and toxicity, and their disposition represents an enormous challenge. As our population continues to increase, so does the amount of waste that needs to be disposed. Some of the waste created by humans is from our own metabolism. In rural areas of the world, human metabolic waste is just buried or dispersed. But in urbanized areas the amount is so large that we must collect and treat our metabolic wastes. Ignoring this problem leads to all kinds of waterborne diseases such as typhoid and cholera, which still kill hundreds of thousands of people a year worldwide. We're going to consider water pollution, in particular, in a future lecture.

Human waste, or feces, is treated in waste treatment plants. This is a combination of chemicals and physical processes that are used to remove the wastes. There are traditionally three stages of wastewater treatment. There's so-called "primary treatment." Here, the solid materials, and feces, and others are physically removed from the water stream, where they might be treated in a different way. The separation includes wastewater, and so-called "gray water," and street runoff that enter into the sewer system. So, now we have this water, and we need to do something with it.

The second step in treatment we call "secondary." Here, we use chemicals and bacteria to break down the waste. In some sewage treatment facilities is an additional scale of treatment that we call "tertiary." And the effluent, or treated water, is usually returned back to the environment in a local stream or river. Many sewage treatment plants ultimately encounter problems that have pretty significant ecological outcomes. One of the big problems that occurs is when there are large amounts of rain. In this instance, the sewage systems cannot accommodate all the excess runoff from the roads and paved surfaces. Here, if you take a look, we have an example of one of these sewage systems in which they are handling both the effluent that's coming out of human habitation, or metabolic wastes, but are also handling the runoff from roads.

If you notice, the system ends up having to deal with huge differences in the amount of total flow. So, on a typical day with little or no rain, the sewage flows into the sewage treatment plant and can be treated by primary, and secondary, and often tertiary, processes.

But what happens during a rain storm when there's a big additional flush of rain? Here, we run into a problem where the treatment system simply becomes overwhelmed, and there ends up being a system to allow that overflow to simply bypass the treatment process. This is where the excess pollution comes from because the sewage treatment facility can only handle so much. Now, when we cause this overflow, often untreated sewage is released along with the combined overflows.

Another issue with combined sewage overflow is the increase of things like hormones and prescription drugs that are being released into the water that were at some point prescription medications for people. This is due, in part, to an increase in prescription drugs and hormone-like mimics, such as drugs for birth control and fertility. Once those things pass through the body, they're back into the ecosystem. These drugs tend to be water soluble, and small amounts are excreted when we urinate. We don't even have a system established yet for treating water for these kinds of challenges. We know that the increase in hormone mimics and other drugs in the water affects the plants and animals that are living near the outflow. It, in fact, represents one of the most important impacts that we know of in amphibian populations, something that we'll talk about later.

Solid waste or garbage is another problem. As our society becomes more consumer and materialistically driven, we have more and more solid waste to deal with. The amount of garbage or trash thrown out each week in the United States is about 3.5 barrels per household, or about 90,000 truckloads each day.

If we take a look at a landfill image, you get a sense of just the extraordinary scale that that works out to be. It's about 240 million tons of garbage generated annually in the United States. Much of this garbage is household waste, and the percentage of disposability has increased over time as we shift to a use-once-throw-away culture. The solid stuff is called "municipal solid waste," or "MSW," for short. Over the past 40 years the amount of municipal solid waste has doubled in the United States. In 1960, Americans generated about 2.5 pounds per day. Currently that's up to about five pounds per day, and is continuing to grow.

The cost of management and disposal of this waste is very high, and the ecological impacts of the trash are absolutely devastating. Even the recycling activities that humans do can cause significant challenges, and often result in significant social inequities that lead to environmental injustice. Historically, cities have either buried their trash, or burned it, or took it out to sea and dumped it. Landfills or dumps are one way to dispose of trash. Landfills generally begin as a hole in the ground, and if they're relatively modern they might be lined with some kind of an impervious material, like plastic. This helps to prevent the toxic materials from leaching into the ground and adding to possible groundwater contamination. Historically, the problems is, of course, that most landfills didn't have that, and so most of the thousands of landfills that remain across North America and Europe, even though they're not being used now, don't have those kinds of linings. The garbage that is there is continuing to leach toxic materials into the soil and into the ground water that's underneath the landfill itself.

When the landfill is built, layers of trash and garbage are covered with soil. The idea is that microorganisms and bacteria in the soil will aid in the decomposition of the trash. Landfills are generally regulated under the Environmental Protection Act with federal regulations, but each state is responsible for enforcing those regulations, and so what you have is a very complex matrix across the United States of the way in which these landfills are regulated. Many problems occur, like soil and water contamination, as well as poor air quality associated with these landfills.

We have a whole new set of problems related to solid waste disposal that is a result of our use of high-tech materials, like computers, and televisions, and flat screen TVs, that we call "E-waste." People living in the United States have a lot of electronic devices: computers, televisions, handheld devices, mobile phones. They are the hallmark of a modern society, but they bring with them a tremendous legacy of waste when we're done with them. It's estimated that 3 billion devices live in American homes. They all wear out. And that's where the real trouble starts because inside the heart of these machines lies a toxic soup of heavy metals and other poisons.

I have here a piece of an old circuit board that came from a computer. It's part of a video card, and for any of you who have ever tinkered with your own computer this is a pretty common sight. Essentially, it is pretty harmless in this current state. The toxic materials in this card are isolated from our use. But this card is made up of a variety of different materials, many of which can be recycled. And so one of the challenges that we face is that when we go to take these cards apart, they will expose some of the toxic materials that are there, and those people who are actually involved in the process of recycling are at relatively high risk. Even when we don't recycle these materials, we create these mountains of old circuit boards buried in computers. As they get crushed under the weight of materials on top of them, they begin to release the toxic materials that, during the period of time that they're used on a daily basis, represent relatively small risks.

When these electronic devices are built, the toxic materials such as lead and mercury are insulated from the outside environment and are very safe for us to use. Remember, most of the really nasty stuff is confined to the circuit boards themselves. But when the devices no longer work, they're taken apart by recyclers, discarded in landfills, or left to pile up in houses and dumpsters. Americans discard approximately 2 million tons of these electronic devices each year. Collectively, we call this junk "E-waste," and much of it ends up in landfills. Lead, which is a brain toxin, is necessary to make the cathode ray tube that lights up the viewing screen and to make connections on circuit boards. That's just one of the toxic chemicals. Mercury, another serious biological toxin, is found in flat panel displays. We talked about mercury when we discussed the Minamata case in Japan. Cadmium, which is a known cancer-causing agent, is found in batteries and on circuit boards. Now, polyvinyl chloride, or PVC, is highly toxic when it's burned, and is the plastic that houses computers and insulates wires. So, although PVC is relatively inert as we use it on a daily basis, at the end of the use cycle for these materials they become highly toxic as they're disposed of. Later in this lecture we're going to talk a little bit more about the high social cost of dealing with E-waste.

Marine debris and dumping is an entirely other kind of problem and practice that has been used for waste disposal. In 1972, the United States Environmental Protection Agency was given the authority to allow limited ocean dumping in the surrounding ocean waters. Currently, it's banned worldwide after debris began washing up on public beaches. This is a huge problem, as dumping is either under-regulated or not regulated at all. It's very hard to catch the disposal of illegal substances like plastic and medical waste. Think about it. If we think that the challenge of managing waste in the terrestrial environment is hard, remember that 70% of the world's surface is covered by ocean. And so ocean dumping is almost impossible to effectively patrol.

One of the great problems with ocean dumping is a little bit different than landfills. So, we think about landfills here, right? We have the problem of putting garbage inside a fixed location. We cover everything up. Now, we do have materials that leach out and gases that might escape, but we pretty much know where the garbage is. In the ocean, when we dump things, they don't necessarily stay where we dump them. They tend to get moved around by the massive current system. So, this one serious problem of ocean dumping of the many is that trash travels with the ocean current. It may be dumped in one place, but it ends up somewhere else.

It collects in areas where the currents are weakest. One example is the Great Pacific Garbage Patch, of which we have an image here. It's double the size of Texas, and the debris washes ashore to nearby islands, including Hawaii. You can see from the image that these large current systems that are located around the world are contributing to the movement and disposition of trash. There are essentially five large major ocean-wide gyres, one in the North Atlantic, one in the South Atlantic, North Pacific, South Pacific, and the Indian Ocean, and they end up serving not only as conveyor belts to climate, which we will talk about in another lecture, but they literally end up as machines moving garbage around the world's ocean. And again, where the gyres are weakest, the garbage tends to collect, and they become these huge submerged islands of trash.

Tires have historically been used for coral reef restoration. Americans generate about 300 million discarded tires per year. One model was to use some of these tires to actually create anchors on which the coral communities could begin to reconstruct themselves. In the 1970s, about 2 million used auto tires were dumped off the Florida coast in an effort to create an artificial coral reef. These tires were lashed together and weighted down with cement to create additional reef habitat that would hopefully encourage more fish, which they have done. But the problem is, over time the tires have broken apart, and as the rubber deteriorates it releases the toxic chemicals that were embedded in the tires, and that actually begins to restrict and reduce the coral growth. The poisons released by the tires themselves become inhospitable. So, you have this very difficult situation now, where the very structure upon which the corals are built is releasing poisons into the coral's environment.

The tires have also moved around due to things like hurricanes and storms, and as the tires have moved around, not only have they damaged the home reef, but they're threatening the remaining natural reefs that exist.

In 2007, in a huge effort, the Coast Guard and the Navy removed tires that covered about 34 aces of ocean floor. It was a huge project. This was done in response to this rising issue. In response to this there was the formation of the National Undersea Research Program. Part of this group's focus is the study of deep-sea biodiversity as it responds to pulses of human disturbance; in this case, toxicity from dumping.

In addition, these organisms and their role in the food web are of concern, as these contaminants can find their way back into human metabolism. The study area is on the East Coast of the United States in temperate waters off the coast of New York City, only about 100 miles from one of the biggest cities in the world. The history of this dump site includes its use as a receiver of wet sewage, some 40 million tons that was deposited over a six-year period of time, beginning in 1986.

The data suggests that the marine ecosystem's biodiversity was being significantly altered, as was the sediment composition which had levels of silver that exceeded a control area's measurement by a factor of 20-fold. Many benthic marine organisms use the sediment as a home, burrowing into the layers and incorporating organic material into their bodies. Ecologically, this was causing a rise in the numbers of creatures, like sea urchins and sea stars, that could ingest the sludge derived from the organic matter.

In a 1993 study that was conducted by Fred Grassle and Paul Snelgrove, working out of both the Memorial University of Newfoundland and Rutgers University, they found that levels of silver appeared to be on the increase as far as 50 nautical miles south of the dump site. This was suggesting that the recovery of the dumpsite had led to changes in other habitats as re-suspended materials were transported to sites south of the dump site. The insight here is that this material continues to be worked into the system, and ends up being found in organisms that you would essentially, at first blush, consider them not to be impacted.

Back to E-waste. Remember, we talked about how the problem is generated by the breakdown of these materials. We've talked about tires, we've talked about other kinds of waste, but we began a few moments ago talking about circuit boards and that challenge. What we know is that E-waste and toxic metals are becoming an increasing problem, especially in poor nations. Electronics like televisions, digital cameras, computers, and cell phones all contain various toxic chemicals and metals that need to be disposed of properly. Most of these products are not biodegradable, and tend to be sent to poor countries to be taken apart, countries like China, India, and Kenya due to lower environmental standards in those countries.

While E-waste accounts for only about 2% of the trash in landfills, it's about 70% of our toxic waste. According to research being conducted by *National Geographic*, the Environmental Protection Agency predicts that in the next few years about 30–40 million computers will be thrown away. Along with the switch to high definition television, there will be an estimated 25 million TVs that are now obsolete and will be disposed of every year. Cell phones represent another E-waste stream, and in 2005 alone, almost 100 million cell phones were discarded. Every one of these electronic materials contains precious metals that can be recycled, but it is very risky to do so.

In India and other countries where E-waste is sent, people rummage through, collecting parts that contain these metals. They melt down the metal and then collect it, and other waste streams remain. They melt down some of the pieces to collect bits of lead and copper. These materials are often melted in the same pots that the family uses to cook their meals. During this process, the air and soil qualities at these dump sites have very, very high levels of carcinogens and air pollutants.

In 2005, the Basel Action Network released a report on digital dumping and abuse on the African continent. It was found that when landfills got too high, villages would just set them aflame. Of course, that releases additional materials into the environment. The residents, of course, complained of fumes and poor air quality. Children and livestock often roamed through these trash piles, and trace amounts of harmful metals ended up in their diets.

A long-term study is being conducted in China right now that's finding that children living near or at these E-waste dumps had significantly higher blood levels of lead than those living farther away. It was also found that the amount of lead found in the blood increased with age. As you can imagine, continued exposure results in higher levels of toxicity. Think of our conversation about biomagnification and about long-lived creatures. Humans sit at the top of the food chain; we're long-lived; we have high metabolic rates; it sets us up to be susceptible to these kinds of challenges.

This is not to impugn the activity of recycling. Recycling is a critically important means for us to reduce our waste. Remember the three Rs of recycling: Reduce, Reuse, and Recycle. Reducing the amount of energy and waste is central to helping the environment. Reusing items and prolonging the life of items can also be a very useful practice. However, when we engage in this it's complicated. The recycling of plastics, metals, and papers are an important practice, but generally, recycling is a fairly complicated cost benefit analysis. Many people argue that the cost of recycling and removal is actually more expensive than the use-and-discard strategies that have been implemented over the past 50 years. But recycling helps reduce the amount that goes into landfills and, therefore, creates more space.

The EPA has also found that overall, recycling reduces carbon emissions by about 49 million metric tons. Since 1980, America has recycled only 10% of its municipal rubbish. Today, that rate is at about 25%, so we've seen a big change between 1980 and the present. In Austria and the Netherlands, those figures reach about 60% or more of their municipal waste.

Recycling in urban areas is even more cost-effective however, and the long-term benefits have more value and further support the need to recycle. One example is Florida's Trash to Energy Program. Across the country about 150 of these combustion facilities are working nationwide and produce, collectively, about 2700 megawatts of energy by burning 33 million tons of municipal solid waste each year.

Florida has the biggest municipal waste-to-energy capacity in the United States. Nearly 20,000 tons are incinerated each day in facilities like the one shown here and generate over 500 megawatts of electricity. Other municipalities are capturing natural gas—methane—that is released from old landfills, and are using the gas to power and heat local communities.

For example, the University of New Hampshire became the first major institution of higher education to tap into the gas produced by landfills. They built a 13-mile pipeline from a local landfill, and now the University of New Hampshire powers about 85% of their on-campus energy needs from this renewable resource. Called "Ecoline," this project lowers the university's carbon emissions, and it stabilized their energy costs by tapping into a sustainable source of fuel.

One way to help ease the impact of E-waste is to find new ways to build, discard, and recycle electronic devices. The architect, William McDonough, is an expert on environmental design, and has written extensively about it. In his book, *Cradle to Cradle,* he suggests that we don't have to be buried in E-waste. Instead, we need to reinvent the way things are built. Using a technique called "Life Cycle Analysis," or LCA, we can assess the environmental impacts of products, along with the energy they use.

If we can make products more safely and find better uses for them after they are no longer needed, the negative environmental impacts will be reduced. One example is synthetic paper that is made from plastics that would otherwise end up in landfills.

As you can see in this context, there is a lot of work to be done, and in our next lecture, we're going to move from a consideration of trash to an understanding of the movement of water. As you know, the trash and water nexus is a pretty tight one.

Stay tuned, and thank you.

Lecture Eleven
The Water Cycle and Climate

Scope:

Water is so essential to life that its role in the ecology of living systems is unparalleled. It is a limiting factor in ecosystems and has a dramatic effect on climate. In this lecture, we investigate the water cycle and the role that it plays in ecosystem services, pollution abatement, and climate.

Outline

I. Water is often the key limiting factor to life within an ecosystem.
   A. About 70% of Earth is covered with water.
   B. Saline water makes up about 97% of all the available water on Earth.
   C. Of the remaining 3% that is fresh water, only about 0.001% is in the atmosphere at any time.
   D. The process by which water enters the atmosphere is the conduit between the large supply of water in the ocean and the freshwater reserves.

II. The fresh water that we can access cycles through the atmosphere and is vital to all living things.
   A. Plants and animals in terrestrial environments have adapted to life based on the amount of water that they can access.
   B. If there is a drought and less plant growth, all consumers and higher trophic levels will experience a shortage in food and available energy.

III. We consider how water interacts with ecosystems by discussing the water, or hydrologic, cycle.
   A. The critical steps are evaporation, condensation, and precipitation.
   B. As the Earth warms, water is heated and evaporates, turning into water vapor in the air.
   C. Most of the evaporation is from the surface of the oceans.

- D. Water vapor rises up into the atmosphere, then cools and condenses to form clouds, which are tiny frozen crystals of water and dust.
- E. Gravity and temperature variation eventually compel the water to fall down as precipitation.
- F. That precipitation will often return to liquid form and flow to bodies of water, beginning the cycle again.

IV. While on Earth, water can remain on the surface of the land and move as sheet flow, or it can percolate into the soil and become part of the great reservoir called ground water.
  - A. The surface and the ground water interact in many ways, and they are actually part of the same system.
  - B. Surface waters are relatively easy to access; ground water is more complicated and in some instances impossible to access.

V. For a given stream or river, all of the land that its water comes from is called its watershed.
  - A. Watersheds are natural ecological boundaries for planning and management.
  - B. Watersheds may consist of dozens of towns, or the boundary between 2 political entities may fall in the middle of a river.
  - C. To address this conflict between ecological and political boundaries, people have created watershed associations.

VI. One of the important aspects of water is its chemistry, which enables it to do a lot for us with respect to ecosystem services.
  - A. Water has high surface tension, adhesion, and cohesion; water also has strong capillary action.
  - B. Water has a very high specific heat, which helps it moderate Earth's climate.
  - C. Areas near water have more moderate climates, so communities placed near water do not get as hot in the summer or as cold in the winter.
  - D. Approximately 40% of the world's population lives within 50 miles of a coast.

VII. Key changes in the hydrologic cycle associated with an increased concentration of greenhouse gases in the atmosphere have resulted in warmer conditions.

Suggested Reading:

Leopold, *A View of the River*.

Wright, *Environmental Science*, chap. 7.

Questions to Consider:
1. How does water cycle through the global ecosystem?
2. What is an ecological watershed, and why is it an appropriate scale of study in ecology?

Lecture Eleven—Transcript
The Water Cycle and Climate

Hello, and welcome back. I hope I didn't frighten anyone off in our conversation about trash and E-waste in the last lecture. I think it's actually quite hopeful, some of the technologies that are emerging. In the last lecture, we focused quite a bit on the interface between trash and its impact on the aquifer of flowing water that runs underground.

You know, water, like all other molecules in ecosystems, is recycled and used over and over. Water is such an essential molecule to life, its role in the ecology of living systems is completely unparalleled. Even something as far out as astro-biology, the search for life in other planetary systems, really can be reduced to a search for water. If astro-biologists, through their technology, can find any evidence of water or water vapor on other planets or moons, that's really a pretty strong indicator that life is possible.

Water is a limiting factor in ecosystems, and it is the common conduit that connects various trophic levels in the biosphere. Water has a dramatic effect on climate. Its physical chemistry makes water a very potent storage system for heat, something we'll be exploring later. We'll investigate the water cycle and the role that it plays in pollution abatement, disease, and long-term sustainability.

I want to begin by talking about a particular species of animal that is adapted to deal with water stress, and that example is the sand plover, which is found in Africa. They nest on substrate; they're not tree-nesting birds. They nest on the ground, and traditionally they nest during the driest season of the year, which has advantages for them with respect to them avoiding predation and that type of thing.

The young chicks hatch out into an environment of pretty high water stress. Although the females do the bulk of the incubation of the eggs, the males have adapted a special set of breast feathers that allow them to collect water. What they do, is they'll fly sometimes 3–5 miles away to the remaining water reserves that exist in the ecosystems where they're nesting. They soak their breast and belly feathers in the water, and they fly back to the nest. When they return, the young chicks literally bury their heads into the breast and belly of the male's feathers, and from that they can take the water out. It's really quite an extraordinary thing to see, and it's quite an adaptation because feathers, for the most part, have evolved to repel water.

Their physical structure literally repels water. It's essentially an organic raincoat for birds to wear. But in this instance, these feathers actually absorb the water. Without this connection, despite the fact that the birds have habitat and are relatively predator-free, they wouldn't be able to complete their life cycle if they couldn't get critical water resources to the young.

Water, like all materials, cycles throughout the ecosystem, and water is often the key limiting factor to life within an ecosystem.

As you know, about 70% of the Earth is actually covered with water. The oceans and seas, which contain the saltier, saline water, comprise about 97% of all the available water on Earth. The other 3% is fresh water, but over 2% of that, about 2.1%, is trapped in ice caps and glaciers, although that figure is going down as glaciers melt. About 0.1% is found in freshwater lakes, and a really small amount, 0.0002%, is found in rivers and streams. An additional amount, although small, is very critical. About 0.001% is actually in the atmosphere at any one time. Typically, water stays about nine days in this atmospheric form, and is the process that creates all of the freshwater resources on the planet.

Stop and think about this for a minute. We have 97% of our water that is salty, and essentially not accessible to terrestrial organisms from a metabolic standpoint. The rest of the fresh water is trapped in ice and glaciers. We have this finite, but renewable, fresh water resource and the conduit between the large standing supply of water that is in the ocean, and freshwater reserves, is that process by which water goes to the atmosphere.

Only small amounts are in the atmosphere at any one time, but that amount is critical, and I'm harping on this because we're going to talk about the role and movement in water vapor in large air currents, both in this lecture and in future lectures. So, the potential for human use to outstrip freshwater supply—we're going to look at that in detail in the next lecture because it actually may represent the single most important environmental crisis we have. We think about the Earth in crisis. Fresh water availability is probably right at the top of the list.

The available fresh water that we can access cycles through the atmosphere and is really vital to all living things. Plants and animals in terrestrial environments have adapted to life based upon the amount of water that they can access. We talked about the sand plover, but certain plants, like cacti living in the desert, have large vacuoles in their cells for water storage, and they carry out photosynthesis at night to reduce the amount of water that they lose.

All plants have a dynamic water balance. During the day, special pores on the undersides of their leaves open to exchange gases. Oxygen flows out, carbon dioxide flows in. However, the gradient for water is steeper than the gradients for the other gases. Inside the leaf is essentially 100% humidity, and when those pores are open, collectively called "stomata," there is extreme water loss as these stomata remain open. At night under more humid conditions, plants close their stomata and regain some of the lost water through uptake.

The amount of available water also influences the producers and the amount of available food. If there is a drought and less plant growth, then all the consumers and trophic levels above will experience a shortage in food and available energy. Remember, when we talked about trophic dynamics, we said that there are controls from the top down, and there are controls from the bottom up. Remember, the top-down controls are things like predators, but the bottom-up control is how much producer metabolism is turned into biomass so that primary consumers can eat them, then secondary consumers, and so forth. Under periods of drought, a smaller portion of producer biomass is generated, which ends up limiting the whole system.

One way for us to consider how water interacts with ecosystems is to concentrate on our understanding of the water cycle. It's really a way that we can understand how water moves through an ecosystem. Also called the "hydrologic cycle," the critical steps are evaporation, condensation, and precipitation. That couples with other transport, both at the surface of the Earth and below.

Just as we considered other materials in the ecosystem, water can be considered to reside in different compartments, only some of which are available to humans and other organisms for use in metabolism. So, as the Earth warms, water is heated and evaporates, turning into water vapor in the air. Most of the evaporation is from the surface of the oceans. However, water also evaporates through both evaporation and transpiration through plants. Oceanic evaporation is the first step in creating fresh water for use by terrestrial and aquatic organisms.

Water can also evaporate, as we said, through the process of transpiration, which we also call "plant sweating." This is a critical factor in plant metabolism, and water in this instance literally serves as a plant's bloodstream, moving materials around the plant.

Later on, when we're going to be considering the role that urbanization plays on landscapes, one of the things that happens, is the amount of vegetation is reduced, and so that evapo-transpiration, that combined process, is profoundly altered when we take plants away from an ecosystem.

We're focusing on the physicality of the water right now, so as the water vapor rises up into the atmosphere it cools and finally condenses to form clouds, which are tiny frozen crystals of water and dust. These clouds, formed by water vapor and dust particles, move throughout our atmosphere, and eventually gravity and temperature variation compel the water to fall down as precipitation. This precipitation can come down as rain, snow, sleet, or even hail.

If the rain lands on a plant leaf, well, it's a pretty short journey, as it will be quickly evaporated back into the atmosphere when the Sun comes out. However, if the precipitation falls onto the surface of the Earth, it will often return to the liquid form and flow to rivers, lakes, oceans, and streams, where it is again warmed by the Sun, and the cycle begins again.

While the precipitation is on the Earth, the water can remain on the surface of the land and move as sheet flow, where it will eventually move into streams and rivers, or it can percolate into the soil and become part of this great reservoir we call "ground water." The surface and the ground water interact in many ways, and they're actually part of the same system. But we divide our understanding of surface and ground water from a functional standpoint because to us, surface waters are relatively easy to access. Ground water is a little bit more complicated, and in some instances, impossible for us to access.

For example, as ground water is pumped out of wells, nearby streams and rivers will actually fall in level as the water table adjusts to its reduced volume of water. Here is something you probably haven't thought of, and there are exceptions, but in general, when you're traveling about and you see streams, and especially when you see lakes, you can imagine that the lake is actually the water table. It's actually the level of water in your area. As you move away from the lake the land form tends to increase in height, and you can imagine, actually, that in a very rough way, the level of water actually stays the same within the ground level. For different geological reasons it might sink or it might rise, and if you're in a community that has seawater around it that's a different story as well. But we can really think of lakes as openings in the water table. That's why there's such an intimate connection between ground water and surface water.

This relationship is particularly acute in coastal communities where the freshwater literally floats on top of the seawater. There's a slight difference in density, and fresh water floats on top. As the freshwater is pumped out of the ground, localized depressions in the aquifer will encourage saltwater to rise up and infiltrate the freshwater lens. We call this "saltwater intrusion," and once that happens, it's very, very hard to re-establish your fresh water supply. This is particularly acute because such a high percentage of the world's population now lives on coastal margins, and we're consuming more and more water from the coastline. Not only are the water tables falling, making it more of a challenge to pull water from the ground, but as the water tables fall, the fresh water lens becomes susceptible to saltwater intrusion, and again, once that occurs, your fresh water supplies are probably permanently compromised.

We consider these systems of water to be lotic because the bodies of water are in constant motion, flowing downhill toward the ocean. This idea of a lotic and moving water system leads us to a conversation about the ecology of the distribution of water in ecosystems, and for that, we need to introduce the idea of a watershed.

Although water is highly mobile, it's useful to consider the natural geologic factors that tend to constrain its movement. When water falls to the Earth as precipitation, it strikes the ground and flows in response to gravity towards a stream and then a river, maybe a lake, and eventually out to the ocean, always flowing downhill. For a given stream or river, all of the land in which its water comes from is called the "watershed." We can think of a watershed as sort of like a big bowl surrounded by mountains, going down into plains, and into lower areas, and eventually into streams, then into rivers, and out to the ocean. So, a watershed, with a river at its basin, all the land that collected the water that is in that river or lake is part of the watershed.

The community that I live in, in the Charles River watershed in Boston, the Charles River is approximately 80 miles in length. It comprises a watershed of about 300 square miles. So, theoretically, and actually, you could trace all the water in the Charles River to somewhere in that watershed. It fell to Earth as precipitation, it moved through either sheet flow or penetration, and became part of the ground water that moved into the Charles River. We can also assume that pollution found in the river must have come from that area, as the water in the river was delivered from that land area.

Watersheds are natural ecological boundaries for planning and management. This is made even more prescient when we consider that an average of 4200 billion gallons per day of fresh water falls on the watersheds of the contiguous United States. This is an important ecological consideration, and when we have talked about this model for understanding ecosystem services, the so-called "ISSE model," we've talked about bringing together the drivers of biogeophysical systems and the drivers of human social systems. When we talk about watersheds, we are at that either confrontational or collegial space. Here's what I mean.

Think about the boundary of a watershed. It makes wonderful ecological sense. The watershed is the land area that captures all of the water that we find in the river. If we want to manage the Charles River water quality, we should be managing from an ecological standpoint all of that watershed. But that watershed consists of dozens of towns. And even more complicated, when we think of the political boundaries that are overlaid on ecological boundaries, they often are a pretty ill fit.

For example, the Charles River divides the cities of Cambridge and Boston. The actual boundary between the cities is right in the middle of the river. That makes perfect sense if you're worrying about taxation or where children should go to school because any house is going to be built either on one side of the river or the other. It's pretty clear who you should be paying your taxes to and where your children should go to school. If you have a fire at your house, what fire department do you call?

From an ecological standpoint, think of what a challenge that is because if your city boundaries are right in the middle of the river, then where is the benefit for one city to manage their pollution, which is very costly, if the other city is not going to do so. It returns us to the tragedy of the commons challenge that we have in ecological restoration and management.

In order for these kinds of political and ecological incongruences to be solved, we actually need a sort of additional contribution to the idea. And so we actually have the emergence of things like watershed associations and watershed management districts to try and get around this problem between an ecological and a political boundary.

Watersheds are particularly useful concepts, as many ecological outcomes of water movement are associated with watershed dynamics. Later, we're going to see how watershed models help ecological resource managers, especially when we talk about urban watersheds.

Watershed dynamics are impacted by local topography, weather conditions, and parent soils. For example, mountain ranges cause upwellings of air that squeeze out the rain on the upwind side of the mountains and create a rain shadow on the downwind side. So, those end up being two separate watersheds, right? You have a set of mountains. On one side of the mountain the water flows toward one set of rivers, and on the other side of the mountain it flows toward another. One side, if this mountain range is intercepting a prevailing wind direction, is going to be a very wet watershed; the other side is going to be a very dry watershed.

In addition, soil and rock characteristics will determine the rate at which rain water infiltrates the ground and percolates through the layers beneath. As humans alter the soils through development, compaction, and agriculture, surface and groundwater relations change, typically in ways that reduce the amount of freshwater available to watersheds. This is critical because as land becomes compacted, it becomes altered. Generally, water flows faster, which means less of it gets into the ground water, which means less of it is available for us as humans. One impact of this is something called "desertification." It's a huge global problem that we're going to focus on in the next lecture.

One of the important aspects of water is its chemistry. It's very interesting, and it enables to understand some of the ways in which water does so much for us with respect to ecosystem services. Most all of you know that the basic structure of water is a central oxygen atom bonded with two hydrogen molecules. But due to the polar nature of the bond, because the electrons involved in the bond are not shared equally between the oxygen and the hydrogen, it creates a charged environment. Due to the positive charge of the hydrogen molecule, and the negative charge of the oxygen, the water molecule forms a particular shape, and this molecule is also considered polar, and it gives it properties that are essential to life.

First of all, water molecules stick together, or bond, through the hydrogen bonds, which also contribute to its ability to have high surface tension, adhesion, and cohesion. Water also has a strong capillary action, which means it will climb up the sides of surfaces, which enables water, for one thing, to move up plant xylem and phloem. It's one of the chemical characteristics of plants that allows movement of water to be possible.

Water also has a very high specific heat, which helps it moderate the Earth's climate, which is a fancy way of saying that you can put a lot of heat, relatively speaking, into a molecule of water before it changes its temperature, which means as it cools down it releases significant amounts of heat.

Water is also a universal solvent. The erosions of valleys and soil transport are critical facts associated with its solvency. It's also typically slightly acidic, and therefore, water solubilizes many chemicals, including salts and metals.

Water is also less dense when frozen, and this has some very interesting implications. Among other things, lakes will freeze from the top down, despite the fact that cold water is more dense than warm water and holds more oxygen. These appear to be disparate statements, so let's take a moment to make sure we understand this.

Typically, as molecules are heated up, their motion increases, and the space between them increases. And water is the same way. As you heat, water it becomes less and less and less dense. As you cool water down, it becomes more and more and more dense and, in fact, holds more oxygen while that is happening, which is why high metabolic demand aquatic organisms, like trout, tend to be found only in fast-moving streams that are cold and have lots of oxygen.

Something very interesting happens to water. As it gets colder, it gets denser and denser, and you know this to be true—I don't need to explain this to you—on a summer day when you walk into a lake to go swimming, the surface waters are warmer, and then your feet are pretty cold in a big lake because the lower waters stay very dense and cold. But something very interesting happens as the hydrogen bonds begin to slow down and become permanent at the moment water forms ice. They take up a little bit more space, and ice now has less density than cold water. And so as a result, on the surfaces of lakes, for instance, ice freezes at the top, and it usually leaves a layer of air between the ice and the water, which serves as insulation. Without that, if water behaved like some of the other chemicals and became denser when frozen, ponds would freeze from the bottom up, and that would have a profound effect in temperate climates in the distribution of life.

Sunlight penetrates water and allows for photosynthesis in aquatic plants. Not all liquids allow sunlight to penetrate, another critical piece of water's chemistry.

One of the most important things that water does is to modulate the climate. Areas near water are more moderate, so communities that are placed near water don't get as hot in the summer, nor as cold in the winter. Remember, water's chemical structure and high specific heat capacity moderate the Earth's climate, especially those areas or cities near a large body of water that will experience these kinds of effects.

Approximately 40% of the world's population lives within 50 miles of the coast. Areas near oceans and large lakes experience temperature differences differently than areas that are not near lakes and oceans. For example, on Cape Cod, a coastal community, during the fall, temperatures remain warmer and milder for longer than inland communities, and the leaves change color later than interior Massachusetts. This is because the surrounding ocean water is warm and helps warm the air.

However, during the spring on Cape Cod, the opposite occurs. The ocean waters are still very cold and take a considerable amount of time and heat to warm up, so the air temperature on Cape Cod during the springtime remains cooler, and it takes longer for springtime to arrive in these coastal communities. This plays out in something we call "lake and ocean effect snow," where warming air masses suddenly get very cold when they hit the water, and then they dump their snow. And so areas around the Great Lakes and areas around the coastal communities are very familiar with what we call "lake effect snow." Buffalo, New York, is a place that's famous for the huge amounts of lake effect snow that it encounters.

According to Andrew Goudie and a team at Oxford University, key changes in the hydrologic cycle associated with an increased concentration of greenhouse gases in the atmosphere has resulted in pretty significant climate changes, including warmer conditions, which increase evaporation rates and result in more intense precipitation events. Over the past 50 years, the U.S. and Canada have experienced 15% overall increases in precipitation rates. That has resulted, also, in changes in the seasonal distribution and the amount of precipitation.

Changes in the balance between snow and rain are pretty significant. Remember, in the winter when precipitation falls as snow, it creates an insulating blanket on the surface of the Earth that prevents additional moisture from being lost from the ground. When that precipitation falls as rain, the snow is melted and removed, and then on dry, cold days significant amounts of moisture are lost from the ground. As climate changes, there is an increased fire risk, and an accelerated melting of glacial ice. That has led to an increased risk of coastal flooding and changes in the dynamic relationship between the growth of plants and the stability of ecosystems.

One of the biggest changes with respect to our understanding of water has to do with our sort of political considerations of how ecosystems store water. One of the changing ideas is how we think about our wetland areas. It wasn't long ago, 100 years or so, when swamps and wetlands were considered a pejorative part of our ecosystem, and we had strategies for draining swamps and filling wetlands. In fact, many of our coastal cities, many of our oldest legacy cities in North America, in fact, have significant portions of their built environments on top of wetlands. This was before we understood how critically important wetlands were in storing floods, in modulating climate, in filtering the water, in being important areas of biodiversity.

One of the interesting parts of my own career as an ecologist has been to see the general public understanding of wetlands change from this pejorative nature of, "We must drain the swamp," to the expectation that wetlands actually do have critical ecosystem services that they provide for us.

In the next lecture, we'll continue this conversation about water, and we'll investigate the impact of water consumption by humans on the ecology of the planet.

Lecture Twelve
Human Water Use and Climate Change

Scope:

In this lecture, we consider the human aspect of the issue of water dynamics. Human population growth is placing increased demands on the freshwater supply. As more water is impounded in reservoirs for human use, more rivers will dry up. The result is a major change in surface hydrology, which will result in the complete collapse of certain ecosystems.

Outline

I. We must examine the overall water budget for a region.
  A. In North America, precipitation either ends up evaporating or forming a relationship that we call runoff.
  B. Runoff is ultimately captured by streams and lakes and makes its way back to the oceans.
  C. Huge reservoirs are available as ground water, but ground water is often hard to access, making it very costly for human consumption.

II. Humans are having global impacts on water supply and distribution.
  A. The average water consumption each day in the United States is approximately 1300 gallons per person.
  B. In Europe, the consumption of water is about half of that, and in certain arid regions of the world, people live on 5 gallons a day or fewer.
  C. There are huge variations worldwide, but typically, usage is about 70% for irrigation, 20% for industrial use, and only about 10% for direct human use.

III. To understand this, we have to work with the ideas of consumptive versus nonconsumptive use and instream versus offstream use.
  A. In consumptive use, water being used by humans is not returned to the ecosystem.

- B. In a nonconsumptive model, the water is used for human needs but stays within the ecosystem.
- C. In instream use, water is used and then returned to the same system.
- D. Offstream use is where water goes back into the aquifer, but in a different area from which it was captured. That is one of the most important aspects of human water use: the diversion of water to an entirely different ecosystem.

IV. Climate change due to humans has increased carbon emissions and is also depleting the amount of water available.
- A. Towns that once depended on snowmelt for water are experiencing fewer available water supplies.
- B. Warmer climates will most likely generate more severe storms.
- C. Climate change will alter the intensity of precipitation, snowmelt, and runoff.

V. There is a direct link between human water use and poverty.
- A. The World Bank reports that 80 countries now have water shortages that threaten health and economies, while 40% of the world has no access to clean water or sanitation.
- B. The World Health Organization has striking figures on the impact of waterborne diseases: Diarrheal diseases now kill 3 million each year, while malaria kills 1.5 million.
- C. The demand for water is doubling every 21 years, and even more rapidly in certain regions.
- D. Developing countries like India and China are seeing drastic water shortages as their growth skyrockets and their populations undergo demographic transition.
- E. A wide range of nations are moving toward a more urbanized demographic distribution of their human population, and this has profound ecological effects.

VI. One of the strategies that people use in response to this increased demand and reduced supply is to divert water from rivers, the so-called offstream model.
- A. Dams have been primarily used for the production of electricity, and decisions were made with only hydropower in mind.

  B. Long-term ecological damage is now being factored into the equation of water management.
  C. Dramatically altering the course of these rivers and the volume of water that they deliver changes the delivery of nutrients downstream, with significant impacts.
  D. This is an ecological conundrum, because dams have tremendous positive potential but also tremendous ecosystem impact.
VII. Water conservation is probably the most important part of creating sustainable water practices.
  A. Water conservation is the preservation of freshwater resources at their source.
  B. Agricultural reform is a key element in both reducing water use and modulating climate change.
  C. Home water use is another important area for positive change: U.S. personal consumption could be drastically reduced without sacrificing quality of life.

Suggested Reading:

Goudie, *The Human Impact on the Natural Environment*, chap. 5.

Reisner, *Cadillac Desert*.

Questions to Consider:

1. What are the global trends in human water consumption?
2. Why is human water use so central to developing sustainable environmental practices?

Lecture Twelve—Transcript
Human Water Use and Climate Change

Hello, and welcome back. We're going to continue our conversation about water today, as we consider the human aspect of the equation with respect to water dynamics. You know, 1.5 billion people live under extraordinary water stress that leaves them vulnerable to disaster at any moment. It's predicted that by 2025, nearly 70% of all the world's accessible fresh water supply will be used each year by humans. As more and more water is impounded in reservoirs for human use, many more rivers will simply dry up. The result is a major change in surface hydrology, which is going to result in the complete collapse of certain ecosystems.

To begin this conversation today, I want to talk about an incident that occurred a few years ago that may not directly sound like a conversation about water, but in reality it is. I was coming back from a meeting, and it was just about sunset, and I was driving along one of the small country roads on Cape Cod that gets very busy in the summer. The traffic all of a sudden ground to a halt. It turned out there was a large female snapping turtle in the middle of the road. She had stopped traffic; she was quite large, and people were wondering what to do. It turns out that it was near a neighbor's house, and she recognized my car, and she came up to me and said, "You need to help this turtle. What do you think we should do?" Well, we had to get a big trash can, and put her in the trash can, and move her to the pond where she was headed.

The issue here is that she was moving from one area of the ecosystem in which she lived to the other, and her movements were completely water dependent because although she spends part of her adult life in the water, she actually has to lay her eggs on soils that are reasonably well drained so that the eggs can breathe, but not dry out. And even though she's a terrestrial animal, the relationship that she has with water, both as an adult and for her eggs, is so incredibly critical. And both aspects of what she does have very specific water-dependent uses. Although we think of this as a metaphor, it's actually a real aspect of her ecology. She needs very different environments, so she was crossing the road to get from the wet watershed area that she was in, the sort of basin of the watershed, over to drier soil, so we helped her do that and got the traffic moving.

When we consider the consumption of water by humans, we must factor into this considerations of the overall water budget for a region. For example, in North America, if we take all the precipitation, it's either going to end up evaporating or forming a relationship that we call "runoff." So, precipitation minus evaporation ultimately equals runoff. And each year about 18,000 cubic kilometers fall on the surface of North America. About 10,000 cubic kilometers are evaporated, and about 8000 cubic kilometers, or 2 quadrillion, or 2 million billion gallons, flow as runoff, ultimately captured by streams and lakes, and make their way back to the oceans.

Huge reservoirs are available as ground water, but often ground water is quite hard to access. There are approximately 5×10^{16} gallons of ground water available—that's 50 million billion gallons—but that ground water can be very difficult to access and very, very costly to bring for human consumption.

Humans are having global impacts on water supply and distribution. That's really one of the core conversations that we're going to be having today. Human population growth is one factor contributing to the shortage of water supply. According to the U.S. Geological Survey, the average water consumption each day in the United States is approximately 1300 gallons per person. That works out to almost 400 trillion gallons of water per day, 142,000 trillion gallons per year.

On a typical day, 164 gallons are used for showers, and laundry, and toilets through direct consumption by people living in the United States. Surprisingly, 485 gallons of that water consumption is used for agriculture. We're going to talk more about that later. Seventy gallons are used for industrial use, and about 580 gallons for the production of electrical power.

There are huge variations worldwide, typically, about 70% for irrigation, 20% for industrial use, only about 10% for direct human use. So when we see calculations of the way people use water, often we're only taking into account that 10% that is used directly, when, in addition, there are huge amounts, even larger amounts of water, that are being used to support the infrastructure and the metabolic needs of humans.

In Europe, the consumption of water is about half, and in certain arid regions of the world, people live on five gallons a day or less. In order for us to really understand this, we have to work with essentially two ideas here. First, we have the idea of consumptive and nonconsumptive use, and second, the idea of instream and offstream use. So let me work with the idea of consumptive and nonconsumptive uses first.

In consumptive use, the idea here is that water being utilized by humans is not returned to the ecosystem. In a nonconsumptive model, the water is used for human needs, but stays within the ecosystem, so for instance, the turning of a water wheel is a nonconsumptive use. The removal of water where it is used for irrigation or something else, and it is lost to the ecosystem, is an example of consumptive use.

In addition, we have instream and offstream use. Here, we're talking about the difference between where the water, when it is returned, where it's returned. So in an instream use we're essentially talking about a loaded flowing system. Water is utilized, and then it's returned to the same system.

Offstream use is where it may end up going back into the aquifer, but not in the area in which it's captured. That's one of the most important aspects of human water use that we're going to be talking about. We're essentially talking about diverting water from one lotic system and having it used in an entirely different ecosystem.

Climate change due to humans has increased carbon emissions, and it's also depleting the amount of water that's available. As the Earth warms, and less snowfall occurs in some areas, towns that once depended on snowmelt for water are experiencing fewer available water supplies. This pattern is part of a positive feedback loop as the melting of snow and ice reveals dark substrate, which in turn warms and hastens further melting of ice and snow. This is why roads in colder climates are dyed black, as it facilitates the melting of ice.

In ecosystems this is pretty dangerous. Instead of the snow cover, as we said before, holding in moisture and because of its albedo effect, reflecting heat, we now have snow melting, exposing the dark surface, adding additional heating, which reduces snow cover even more dramatically.

We also know that warmer climates will most likely generate more storms that are severe, as more energy is available to the systems as storms form. Think about this. We mentioned before in a previous lecture that one measure of heat is the movement of molecules. If we add more energy to the system in the form of climate warming, then there's more energy available as these storms form.

The Millennium Ecosystem Assessment made it quite clear that climate change will alter the intensity of precipitation, snowmelt, and runoff. Loss of surface water and disrupted soils lead to desertification of impacted lands, especially those that are grassland ecosystems.

We have had a significant environmental wakeup call if we think about the impact of human water use globally, and there's no more important lesson we can learn than that from the Aral Sea. It's one of the world's largest inland bodies of water. It's a saltwater lake located in the countries of Kazakhstan and Uzbekistan. During the Soviet era, central planning gave birth to the idea of a cotton-growing region that would be fed by the two main rivers that feed the lake, the Amu Darya and the Syr Darya. Water diverted from the two main rivers that fed the lake has caused massive ecological destruction as the result of lower water levels. Remember, as we said, lakes and rivers represent the tops of the water table. If we're using this water in an offstream model, and we begin to lower the water table, it has profound effects.

The surface area of the Aral Sea has decreased by 90%, and the volume by over 50%. This has caused an increase in salinity, and it destroyed the sturgeon fisheries that were the original attractant of populations to that region. As a result, the tourism industry has collapsed, and it really has been an environmental disaster. A once thriving ecosystem with vibrant marshes is now ringed with salt flats that don't support the previous biotic community. Dust from the shoreline has created a serious air pollution problem, and the climate has become much more severe.

With the collapse of the Soviet Union, the five surrounding countries, with the help of the World Bank, have launched an $85 million project to restore a portion of the lake that is isolated from the rest. A dam has been constructed that allowed the smaller portion to fill with fresh water and bring back the former ecosystem, at least in that pocket of the lake. Overflow from the dam is now feeding the rest of the lake, but not enough, however, to re-establish the previous ecosystem. However, at least that small portion the lake is beginning to return some of its original ecosystem function. The only way to return the ecosystem-wide function of that lake would be to stop borrowing water from those two rivers that feed into the Aral Sea, and that's unlikely to happen. But at least the efforts are being implemented to have a small portion of the lake recovered.

There is a direct link with respect to human water use and poverty, and we talk about this in the context of the nexus of water stress and poverty. The World Bank reports that 80 countries now have water shortages that threaten health and economies, while 40% of the world, which represents 2 billion people, has no access to clean water or sanitation. The lack of safe drinking water has profound health implications. Now, data from the World Health Organization is really quite striking with respect to the impact of waterborne diseases. Diarrheal diseases now impact a billion people a year, and kill 3 million each year. Intestinal helminthes, or parasites, kill 100,000 people a year, and impact almost a billion people as well. Schistosomiasis, which is another waterborne parasite, impacts 200 million people each year, and ends up killing 200,000.

One of the most devastating waterborne diseases is malaria; 400 million people each year get malaria, and it kills 1.5 million people. Over half of them are children. Dengue fever, another waterborne disease, impacts 1.7 million people on an annual basis that kills over 20,000. And Trypanosomiasis sickens 275,000 each year, 130,000 people dying. And so-called "river blindness" impacts 17 million people each year and kills 40,000.

According to the World Bank, the demand for water is doubling every 21 years, and even more so in certain regions. Developing countries like India and China are seeing drastic water shortages as their growth skyrockets and their populations undergo demographic transition. Think about the challenge that's happening in these nations. First of all, India and China are two of the most populous nations in the world, and although there have been some interventions with respect to their population growth rates, those countries still continue to grow at rapid rates. Their fertility levels are dropping, but there are so many young people who are reaching adult age, the populations continue to grow. Remember our conversations around population growth.

In addition, these populations are shifting to a different model of consumption. Those nations are beginning to have populations that want to consume as we do in the West, and so their per-capita use of water resources is going up.

China is shifting from an agricultural-based economy to a manufacturing base, and more people are moving into cities. More people in cities means that there is a higher demand for water in concentrated areas, which puts a large stress on their ability to deliver sanitary, potable water. The quality of water is, therefore, deteriorating because of the increased pollution caused by higher densities of people without the appropriate infrastructure available for support of clean water supplies. So, this cycle ends up aggravating existing water shortages. Other nations, poor nations like Nicaragua, will now have even a reduced amount of potable water available because of this intensity of cycle that happens in urbanized areas.

Let's review this for just a moment so you understand because we've talked about a wide range of nations from the most populous ones in the world like China, which is emerging as an economic power, to developing nations like Nicaragua in Central America, which are still struggling to find an economic footing. All of them are moving toward a more urbanized demographic distribution of their human population, and this has profound ecological effects. We end up with a situation where higher densities of people are putting more demands on an already overtaxed water supply system, and the water supply system is being impacted by the fact that increased amounts of sewage are not being treated appropriately. That sewage and these waterborne diseases are leaking back into the water supply, and the water supply then is no longer safe. So we end up with a diminishing water supply, increasing demand, and more stress within the infrastructure. It's a recipe for ecological disaster.

One of the strategies that people use in response to this increased demand and reduced supply is to think about diverting water, the so-called "offstream model." Let's take water from a river, move it somewhere else, and we can use it. Let's think about this. Let's think about the ecology of this diverted water, and we're going to focus on the Colorado River as part of our conversation.

Worldwide, nearly 45,000 large dams, those that are more than 50 feet high, have been built. At least 3000 of these dams hold back in excess of 25 billion gallons of water each. Collectively, they flood

120 million acres of land, and they store 1500 cubic miles of water. Now, these large dams have enormous ecological and social impact. Nearly 40 million people have been displaced by the construction of such dams, and the wetlands, once nourished by the natural floodings of these river basins, are now dry. For instance, of the salmon that try to spawn on the once rich Columbia and Snake Rivers in the western United States, 95% suffer premature mortality as they try to negotiate the dams and obstructions.

The use of these dams has been primarily directed toward the production of electricity, and the values of damming were calculated only with the hydropower in mind. Long-term ecological damage is now being factored into the equation of water management. The Colorado River is an excellent example of water use and humans having a lasting impact.

The Colorado River starts in Rocky Mountain National Park. It flows through Colorado, Utah, Nevada, Arizona, California, and, finally, Mexico. Prior to diversion and construction of the Glen Canyon and Hoover Dams, it used to flow into the Gulf of California. Now, because of the holdback of the water caused by the dams, the river rarely flows through Mexico, and when it does, it's often polluted and very saline.

It was established that those who discovered and used the water source first had legal rights over the use. This is an interesting public-private conflict that arises when water becomes rare. This has become a tremendous basis for tension as the Colorado waters have been diverted and used throughout the years. Diversions to the Imperial Valley in California have nearly dried up the Colorado River delta in Mexico. So what we have is a situation where upstream users of the water are gaining an ecological benefit, and downstream users are suffering because the volume of water being delivered to those communities is drastically reduced.

Other major aquatic ecosystems have suffered, too, and altered local climate as well. The Rio Grande River, one of America's mightiest rivers, no longer flows to the sea, as most of the water is diverted for agricultural uses. The river no longer brings rich sediments to the Delta to support that nearshore ecosystem. So, what you have here is a rapidly escalating feedback system that leads to ecological decline.

The very fact that rivers flood and bring silt helps to create wetlands and the alluvial fans of nutrients that make for rich nearshore ecosystems, build marsh environments, and help protect the mainland from storms and flooding. By altering dramatically the course of these rivers and the volume of water that they deliver, it's changing the delivery of these nutrients downstream that have significant impacts. We can, as a society, manage some of this through technology, but the technology to manage this is so expensive, many feel it would be better off to have allowed those natural ecosystem services to have been provided, but undoing some of these things is pretty difficult. Damming rivers is a controversial strategy for our quest to achieve sustainable power. This flowing, or lotic, water has powered human industry for thousands of years, going back to the earliest waterwheels in Mesopotamia. At the current rate of world industrial expansion, 260 new dams come online each year, which is an actual reduction from the 1000 plus that were being built in the early 1900s.

No conversation about dams would be complete without talking about the world's largest dam project, which is the Three Gorges Dam, which is under construction in China. It has two purposes, really. It's designed to both control flooding along the Yangtze River and to generate electrical power. This project has huge impacts on the biogeophysical and social landscapes of that region. We consider it a mega-driver of ecosystem change, both short and long-pulse duration.

Flooding along the Yangtze was legendary, and nearly 300,000 people have lost their lives in floods during the 20th century alone. The dam, when it's fully complete, will create a reservoir 370 miles long. It will generate 22,000 megawatts of electricity at peak capacity from 32 main generators, plus two more that are used to run the plant itself. But in order to build this, over 1.2 million people were moved, and entire cities are being flooded by the process. The ecological costs of this project are great, and mounting, as the weight of the impounded water appears to have geologic-scale impact, and may actually increase earthquake behavior in the area.

When completed, the reservoir will cover nearly 250 square miles at a depth of about 500 feet. The dam will hold back approximately 10 cubic miles of water. However, as we mentioned before in these other river systems, along with this water comes 500 million tons of silt delivered annually, some of which gets clogged behind the dam.

That silt would normally nourish downstream ecosystems, and is denied to them by the action of the dam. But these are complicated systems. The upside of this is that the power produced from the water turbines can offset 30 million tons of coal per year, which in turn, offsets 100 million tons of greenhouse gases.

So when we think about the solution space that we move to with these ecological conundrums, something like a dam has tremendous positive potential, but also tremendous challenges with respect to long-term ecosystem impact.

We can't really talk about human water use without addressing the notion of water conservation. It's probably the most important part of creating sustainable water practices. Water conservation is the preservation of freshwater resources at its source. It requires a switch in philosophy from a historic approach to water management of how much water do we need, to a more ecologically manageable perspective of how much water do we have and how do we best use it?

The most likely candidate is agricultural reform. It's a key element in both reducing water use and modulating climate change. Forty percent of the food grown in the world must be artificially irrigated, and nearly half of all irrigation water is lost to evaporation and runoff. Now, drip systems and computer-controlled surge systems are a key to transforming the kind of water largesse that we have used with flood irrigation. Studies indicate that drip irrigation not only saves water, but can actually increase crop yield by 10 or 15%. However, with the exception of Israel and Australia, both water-starved regions, most of the world is still using flood irrigation methods, and that's about 97% of the irrigation in the United States.

The cost for conversion to drip systems is expensive. It's can be $1000 per acre. Under current conditions, the real cost of water is borne by the government through construction projects and subsidies. This is the classic tragedy-of-the-commons scenario that we visited earlier in the course. But as water becomes more expensive to the end consumer, existing and newer technologies await the adopters.

Home water is another important area for positive change. The U.S. water use per capita now exceeds about 1300 gallons per day among the most affluent if all the uses are factored in. Many of the world's people live in arid areas on five gallons per day or less. U.S. personal and supplemental consumption could be drastically reduced without sacrificing our quality of life. One example includes Costa Rica. The country realized early on in their protection of natural resources that fresh water was going to be vital to their country. Costa Rica was proactive in protecting and establishing ecological reserves that contain their water resources. The Children's Eternal Rainforest in Monteverde, for instance, is 54,000 acres that protects a large part of the region's most important watersheds. Not only does it preserve the water source, but they've established ecotourism that keeps the visitors concentrated to one small part of the reserve so the rest can remain untouched. This practice has reserved six of the 12 critical life zones in Costa Rica, and helps preserve the plants and animals that live in these areas.

One of the challenges in dealing with water is this public-private interface that we've made reference to a couple of times to here in the lecture. In one of my roles as an ecologist I served on a regional regulatory board that manages the wetland resources, and one of the biggest challenges we had, for instance, was the placement of docks and piers along public water bodies. Here we have the idea that people who have bought land next to a body of water want to be able to put a dock from their land out into the water so they can use their boats more easily. It's an interesting dilemma because the public actually owns the water space, but people own the land next to it. It's an extraordinary struggle between these two systems.

In fact, for instance, in Massachusetts, if you put a dock into a public water sheet, you actually have to provide passageway for people walking along the shore so they can get past your dock without impeding their activities.

In our next lecture, we're going to tackle the forces that shape climate change overall, and you can bet that water is a key player in this process.

So until then, thank you very much.

Timeline

1838	Pierre-François Verhulst models continuously growing population.
1845	Henry David Thoreau moves to Walden Pond.
1854	John Snow links cholera to water contamination.
1859	Charles Darwin publishes *On the Origin of Species*.
1864	George Perkins Marsh publishes *Man and Nature*; Yosemite designated a California state park, later becoming a national park (1890).
1866	Ernst Haeckel coins term "ecology."
1868	Ernst Haeckel publishes *History of Creation*.
1872	U.S. Congress designates Yellowstone first U.S. National Park.
1875	Edward Suess coins term "biosphere" in *The Face of the Earth*.
1877	Karl Mobius coins "biocenose" idea of organisms interacting within a single habitat.
1886	Audubon Society founded.
1892	John Muir founds Sierra Club.
1898	Gifford Pinchot made first head of USDA Forest Service.
1899	Henry Chandler Cowles establishes ecological succession.

1900	Frank Chapman conducts first Christmas bird count.
1908	Godfrey Hardy and Wilhelm Weinberg publish equilibrium model of a nonevolving population.
1913	*Journal of Ecology* founded; British Ecological Society founded.
1915	Ecological Society of America founded.
1916	Frederick Clements publishes *Plant Succession*.
1920	August Thienemann introduces concept of trophic feeding levels.
1926	Vito Volterra publishes model for interactions between 2 species.
1927	Charles Sutherland Elton's *Animal Ecology* introduces ecological niches.
1935	Arthur Tansley coins term "ecosystem," a concept combining the biological community with its physical environment.
1949	Aldo Leopold's *Sand County Almanac* published.
1953	Eugene Odum publishes first textbook, *Fundamentals of Ecology*.
1957	G. E. Hutchinson publishes idea of ecological niche.
1962	Rachel Carson publishes *Silent Spring*.
1964	U.S. Congress passes Wilderness Act.
1969	Santa Barbara oil spill.

1970	First Earth Day (April 22), following an idea of Senator Gaylord Nelson; Clean Air Act signed by Richard Nixon; U.S. Environmental Protection Agency created.
1973	U.S. Congress passes Endangered Species Act.
1975	E. O. Wilson publishes *Sociobiology*; Zygmunt Plater petitions U.S. Department of Interior for protection of snail darter under Endangered Species Act.
1978	President Carter declares state of emergency at Love Canal in Niagara Falls, NY; First International Conference on Conservation Biology at UC San Diego.
1979	James Lovelock publishes Gaia hypothesis; Three Mile Island nuclear reactor accident.
1980	National Science Foundation's Long Term Ecological Research Program founded.
1981	Paul and Anne Ehrlich's *Extinction* examines how loss of biodiversity affects ecosystem services.
1985	Society for Conservation Biology founded.
1986	Fire and meltdown at Chernobyl nuclear reactor.
1988	Intergovernmental Panel on Climate Change established; Norman Myers originates concept of "biodiversity hotspots" (habitats with rapid loss of endemic species).

1992	Earth Summit in Rio de Janeiro produces Convention on Biological Diversity and Framework Convention on Climate Change.
1997	Kyoto Protocol adopted for reduction of greenhouse gases.
2005	Kyoto Protocol in effect.

Glossary

autotrophs: Producers that make their own food through the process of photosynthesis or chemosynthesis.

behavioral ecology: The quantitative study of animal behavior from a systems approach through the lens of evolution.

benthic: The ecological community at the bottom of the ocean.

biogeophysical forces: The nonliving components of a dynamic ecosystem, such as climate, precipitation, tides, and wind.

biomagnification: The process by which the concentration of materials, often toxic, increases as they pass through the levels of a food web.

biomass: The total weight of all living organisms in a given ecosystem.

biome: A large regional ecosystem, such as a desert, tundra, or rainforest.

biotic structure: The composition and organization of the species that live in a particular ecosystem.

coevolution: Mutual long-term impacts on the genetic structure of a population as influenced by another species living in close contact.

community ecology: The study of the interactions among different species living in the same ecosystem—often the scale at which biodiversity and coevolution are studied.

conservation biology: A branch of the life sciences that focuses on the restoration and preservation of rare natural living resources.

demographic transition: The change in family size and reproductive strategy displayed by a species subjected to geographic, social, or resource changes, typically applied to human societies as they urbanize and have fewer children, despite increased wealth.

detritivores: Decomposers such as fungi and prokaryotes, which get energy from detritus (nonliving organic material); they play an important role in material cycling.

disturbance: A force that changes an ecosystem, such as fire, flood, deforestation, or disease; some forms of disturbance are central to ecological resiliency.

ecological carrying capacity: The theoretical limit to the number of individuals of a given species that can live in an ecosystem. Devilishly hard to measure, but the idea serves as a foundation for establishing research questions.

ecology: The examination of the interactions between the biotic (living things) and the abiotic (nonliving things).

ecosystem: An integrated unit of study in ecology that includes all of the biotic and abiotic components inherent in that system, such as a pond, forest, or intertidal area.

ecosystem services: The outcomes of the living organisms and their environments that are mutually reinforcing to the survival of the ecosystem in which they live.

endemic: Species that are found only in a very small geographic range.

eusociality: A relatively rare reproductive system in highly social animals where only a small percentage of adults breed but most adults care for the young. It is most commonly encountered in the social hymenoptera (bees, ants, wasps), termites, and some mammals such as naked mole rats.

evenness: The degree to which different species have relatively similar representation within a community or ecosystem. Often measured in combination with richness.

habitat: The physical environment in which a species lives.

heterotrophs: Consumers at trophic levels above the primary producers and dependent on the photosynthetic output of primary producers. Heterotrophs can be divided into primary consumers (herbivores that eat primary producers), secondary consumers (carnivores that eat herbivores), and tertiary consumers (carnivores that eat secondary consumers—i.e., carnivores that eat carnivores).

hypoxia: In aquatic ecosystems, the dramatic decline in available free oxygen, typically the result of excess nutrients delivered to the ecosystem by pollution, resulting in excessive growth of planktonic populations.

invasive species: A species found in a geographic area different from the one in which it is normally found. Also known as exotic or nonnative species.

keystone species: A species with a disproportionately high influence on the function of an ecosystem, such as a beaver on a pond or humans on a city.

life cycle: The complete process by which an animal is born, matures, reproduces, and dies, which varies considerably among organisms.

microevolution: Genetic changes within a given species that lead to adaptation or even speciation.

phenology: The study of the timing of natural events, such as the emergence of leaves on trees, the start of animal breeding cycles, and the timing of migrations. Phenology has particular relevance with the discovery of global climate change and the impacts that warming has on the Earth's seasons.

presses and pulses: Long-term and short-term disturbances, respectively, that are the prominent drivers of change within an ecosystem.

population ecology: The study of the growth and life-history patterns of an individual species within an ecosystem.

richness: The number of species within a community or ecosystem. Often measured in combination with evenness.

Shannon-Weaver index: A statistical measure of biodiversity that favors an even distribution in the number of individuals from each species encountered.

social ecology: The study of the social patterns of humans in different ecosystems.

systems ecology: The study of the complex interactions among the biotic and abiotic components of an ecosystem, including the influence of humans.

taxon: A closely related group of organisms, such as songbirds or wolves, coyotes, and foxes.

trophic pyramid: A model used to examine energy flow in an ecosystem—often better thought of as a food web. The pyramid analogy emerges from the biomass distribution of most ecosystems that results in most of the biomass found in the producers and increasingly smaller amounts in the primary, secondary, and tertiary consumer levels.

urban ecology: The study of the structure and function of urban areas as ecosystems that includes humans as integral parts and focuses on sustainable and just practices that promote human survival and well-being.

Biographical Notes

Carson, Rachel (1907–1964): American biologist and author of *Silent Spring* (1962). Carson is best remembered for her pioneering work in the examination of environmental damage caused by pesticides, which originally appeared in *The New Yorker*. She worked for the U.S. Fish and Wildlife Service (1936–1952) before turning to writing full time. She became a very successful nature writer, author of such books as *The Sea Around Us* (1952) and *The Edge of the Sea* (1955). *Silent Spring* sent alarm bells ringing throughout the nation as the adverse impact of pesticides, especially DDT, were being uncovered. Her book was a bold step forward for the study of ecology and for women's place at the table of science. Carson died at the age of 57 from breast cancer.

Cowles, Henry Chandler (1869–1939): American botanist studied plant succession (later generalized as ecological succession) along the southern shore of Lake Michigan, later published as *Vegetation of Sand Dunes of Lake Michigan* (1899). While a graduate student, Chandler learned Danish so he could read the original texts of Eugenius Warming. As a professor of botany at the University of Chicago, he was best known for creating the combination of intense fieldwork and academic study that trained the initial generation of American ecologists.

Darwin, Charles (1809–1882): English naturalist who would change the whole nature of biological studies by championing natural selection as the agent of evolution. After nearly 2 decades of research and analysis, Darwin published *On the Origin of Species by Means of Natural Selection* (1859). This volume presented evidence and argued that modern species descended from ancestral species in an unbroken lineage. His concept of natural selection focused on evolutionary change in populations where favorable, heritable traits become more common and organisms that survive pass on those traits to their offspring, with variation among members of a population increasing and characteristics changing. Natural selection is the force that shapes the characteristics of populations, favoring the most reproductively successful combinations, and thus certain individuals.

Haeckel, Ernst (1834–1919): Zoologist specializing in comparative anatomy who coined the word "ecology" (which comes from *oikos*, or "house") in 1866. Haeckel supported the idea that an individual ontogeny maps its entire species history, or phylogeny. He dabbled in the study of human evolution as an aggressive promoter of Charles Darwin, yet he was also a believer in Jean-Baptiste Lamarck's theory of acquired inheritance. He was noted for making wild predictions without any evidence, such as predicting that humans evolved in the East Indies, and he wrote several speculative and popular books, including *History of Creation* (1868).

Leopold, Aldo (1887–1948): U.S. Forest Service researcher and University of Wisconsin professor who worked to forge a union between ecology and conservation and made major contributions toward soil conservation in Wisconsin, leading to the state's soil conservation act. Leopold's notion of a land ethic emphasized relationships between *Homo sapiens* and the living Earth in which we are cohabitants. He cofounded The Wilderness Society (1935), and his posthumously published *Sand County Almanac* (1949) stands as a sacred text for any student of ecology who wishes to understand the importance of conservation. Leopold helped create a practical, citizen-based connection to ecology. His community of service were farmers and hunters, and his ideas of conservation helped protect the soils and populations of game species.

Malthus, Thomas (1766–1834): English clergyman and fledgling economist. Malthus studied mathematics at Cambridge and developed a mathematical model for understanding populations, in which he calculated that the Earth's human population would double every 25 years without population controls. His repeatedly revised and expanded *Essay on the Principal of Population* gave rise to the notion of the so-called Malthusian catastrophe. He influenced Charles Darwin and Pierre-François Verhulst, who developed his logistic model of population growth after Malthus. Remembered in social history for his relatively cruel philosophies, Malthus influenced British government policy, including policy that contributed to the Irish potato famine. His pessimism contributed to the view of economics as the "dismal science."

Marsh, George Perkins (1801–1882): American lawyer, linguist, diplomat, and conservationist. A native of Vermont, Perkins was influenced by the conservation ideas of John Quincy Adams and used the early years of his position as U.S. ambassador to Italy (1861–1882) to write *Man and Nature, or Physical Geography as Modified by Human Action* (1864). An avid forester, he was offended by the deforestation occurring wherever he visited, including the Mediterranean, which he viewed as particularly devastating. He was instrumental in the early conservation movement in the United States and helped to establish the great swath of protected lands in the Adirondack Mountains of New York that would one day serve to protect the drinking water of New York City.

Odum, Eugene (1913–2002): One of the first scientists to study ecology on a larger scale and author of the first and only textbook in the field for decades, *Fundamentals of Ecology* (1953). At the University of Georgia for his entire career, Odum was a steadfastly critical thinker who helped to sharpen the rigor of ecological theory and to separate the politics of environment from the science of ecology. His public persona grew dramatically when he became a featured intellectual at the first Earth Day in 1970. He used ecological models to help reach environmental solutions with the 1998 publication of *Ecological Vignettes: Ecological Approaches to Dealing with Human Predicaments*.

Shelford, Victor (1877–1968): American animal ecologist and zoologist who did most of his work at University of Illinois. A student of Henry Cowles, with whom he helped to found the Ecological Society of America in 1915, Shelford was in charge of the Illinois Natural History Survey from 1914 to 1929. Shelford is best noted for his law of tolerance, which states that distribution of a species is determined by the environmental factor for which the species has the narrowest of tolerance, such as salinity, oxygen, and pH.

Tansley, Arthur (1871–1955): English botanist who founded the British Ecological Society (1913) and championed the idea of ecosystems as an interaction of biogeophysical forces. Although heavily influenced by Eugenius Warming, Tansley rejected Lamarckism. He did important research on species range and the effect of environmental conditions, publishing a huge map of the vegetation on the British Isles (1939) looked at from the ecosystem level. He founded the *Journal of Ecology* in 1913 and was knighted in 1950. In later life, he branched into psychology.

Verhulst, Pierre-François (1804–1849): Belgian mathematician and doctor whose equation for human population growth (1838) became a cornerstone for population ecology. The theory he developed lay dormant until the early 20th century, when researchers began to use his equations to understand the behavior of populations growing under various conditions. These tools remain at the heart of mathematical ecology and are used by such modern luminaries as Sir Robert May and Hal Caswell.

Warming, Eugenius (1841–1919): Danish follower of Jean-Baptiste Lamarck who also incorporated some of Darwin's ideas about natural selection. Warming helped turn ecology into a science through his work in botany, studying the direct descent of living organisms in the same species. He is noted for an important textbook on plant ecology published in 1895 (*Plantesamfund*; translated in 1909 as *Oecology of Plants*) and for pushing the envelope with respect to an integrated approach to living organisms. He is often credited with being the true founding force in ecology.

Wilson, E. O. (b. 1929): A multifaceted life scientist who helped to develop and shape the fields of ecology, conservation biology, animal behavior, and human ecology. One of the world's leading authorities on the ecology of ants (myrmecology), Wilson has made major contributions to the field of ecology through influential works such as *The Theory of Island Biogeography* (1967) with Robert MacArthur, *Sociobiology* (1975), the Pulitzer Prize–winning *On Human Nature* (1979), *Biophilia* (1974), and *The Creation* (2006). From his laboratory at Harvard University, he has shaped the discipline and launched a whole generation of young researchers into the field. His many accolades include the Crafoord Prize, the U.S. National Medal of Science, a second Pulitzer for *The Ants* (1991), and the TED Prize (2007). It will take the next generation to fully evaluate the impact of his work on the field of ecology and conservation.

Bibliography

Adams, C. E., K. J. Lindsey, and S. J. Ash, eds. *Urban Wildlife Management*. Boca Raton, FL: Taylor & Francis, 2006. This text lays out urban wildlife issues from the objects-to-be-controlled paradigm. Many communities subscribe to this approach, but it generally ignores the ecosystem services of wildlife.

Alberti, M. *Advances in Urban Ecology: Integrating Humans and Ecological Processes in Urban Ecosystems*. New York: Springer, 2008. Aberti has made one of the first efforts at unifying this dynamic and crucial science. This work will shape the activities of scholars for a generation.

Alcock, J. *Animal Behavior*. 8th ed. Sunderland, MA: Sinauer Press, 2005. The benchmark for textbooks about animal behavior, Alcock's book links ecology and behavior in a great study of community-level interactions of animals in their physical and social landscape.

Bernard, T., and J. Young. *The Ecology of Hope: Communities Collaborate for Sustainability*. Gabriola Island, BC: New Society, 2008. Bernard and Young chronicle an emerging sense of possibility in communities committed to meaningful change toward sustainability.

Beston, H. *The Outermost House: A Year of Life on the Great Beach of Cape Cod*. New York: Henry Holt, 1928. This is a classic saga of exploration and discovery of the landscape, reinforcing the sense of man's impermanence and frailty.

Cain, M. L., W. D. Bowman, and S. D. Hacker. *Ecology*. Sunderland, MA: Sinauer Associates, 2008. One of the best of the new breed of ecology textbooks that integrates across disciplines and scales. This book can serve as a reference for the course.

Campbell, L., and A. Wiesen, eds. *Restorative Commons: Creating Health and Well-Being through Urban Landscapes*. Newtown Square, PA: USDA Forest Service, 2009. This beautifully illustrated book probes the theory of green landscape as a critical social and physical milieu for healthy neighborhoods.

Carson, R. *Silent Spring*. New York: Houghton Mifflin, 1962. This watershed volume changed our understanding of ecology and the role of scientists forever. Highly controversial, the book helped to elevate our understanding of pesticides and proved that science can be an act of advocacy.

Collins, S. L., and L. L. Wallace. *Fire in North American Tallgrass Prairies*. Norman: University of Oklahoma Press, 1990. Collins and Wallace developed a deep understanding of the ecosystem services provided by fire in grassland ecosystems that stands a model for analysis of ecological presses and pulses.

Darwin, C. *On the Origin of Species by Means of Natural Selection*. London: John Murray, 1859. Darwin's *Origin* stands as one of the greatest works in the canon of the life sciences.

Dawkins, R. *The Selfish Gene*. New ed. Oxford: Oxford University Press, 1989. Dawkins threw a bomb into the argument on the scale of evolutionary analysis with this volume, which was first published in 1976. He has written a slew of great books since this one, but one should start with the 1989 edition of *The Selfish Gene* as it includes rebuttals of his critics.

Destefano, S. *Coyote at the Kitchen Door: Living With Wildlife*. Cambridge, MA: Harvard University Press, forthcoming. Both science and personal experience fill this volume as it captures both the ecology of the suburban landscape and the passion of the author as a scientist and adventurer.

Diamond, J. *Guns, Germs, and Steel: The Fates of Human Societies*. New York: Norton, 1997. Diamond crossed many traditional boundaries of scholarship to investigate the effect of human culture on ecological processes, as well as the role ecology plays in shaping human history.

Forman, R., D. Sperling, J. Bissonette, A. Clevenger, C. Cutshall, V. Dale, L. Fahrig, R. France, C. Goldman, K. Heanue, J. Jones, F. Swanson, T. Turrentine, and T. Winter. *Road Ecology: Science and Solutions*. Washington, DC: Island Press, 2003. A groundbreaking consideration of roads as ecological entities, both as barriers and as conduits of animal movement and territoriality.

Friedman, T. *The World is Flat: A Brief History of the 21st Century*. New York: Farrar, Straus and Giroux, 2005. An insightful review of the world's history through a blend of systems thinking and ecological narrative.

Gonick, L., and A. Outwater. *The Cartoon Guide to the Environment*. New York: Harper Resource, 1996. This small volume is a great way to engage the core ideas of the major environmental issues in an informal but captivating format. This cartoon guide is for real and gathers the nuggets of environmental science in an easy-to-digest offering.

Gore, A. *An Inconvenient Truth: The Planetary Emergency of Global Warming and What We Can Do About It*. New York: Rodale, 2006. This book ignited a national firestorm and started a political and cultural revolution with respect to the understanding of environmental decline and sustainability.

Goudie, A. *The Human Impact on the Natural Environment*. Cambridge: MIT Press, 2000. This is a wonderful resource book for an overview of humans as drivers of ecosystem dynamics.

Grant, P. *Ecology and Evolution of Darwin's Finches*. Princeton, NJ: Princeton University Press, 1996. This long-running study of the Galapagos finches and their evolution by Peter and Rosemary Grant is a benchmark of great science and offers insight into the complexity of ecology's role in shaping microevolution.

Hage, J. *An Entangled Bank: The Origins of Ecosystem Ecology*. New Brunswick, NJ: Rutgers University Press, 1992. This book is a great review of the forces that converged to launch the ecosystem view of ecology.

Halpern, S. *Four Wings and a Prayer: Caught in the Mystery of the Monarch Butterfly*. New York: Pantheon, 2001. This wonderful account of the ecology of migratory butterflies and the people who study them is a captivating read. The ecology of this species is beautiful, complex and a career of study for some of the best minds in ecology.

Kennedy, D., and the American Association for the Advancement of Science. *Science Magazine's State of the Planet*. Washington, DC: Island Press, 2008. This striking report first appeared in the Nation's top peer-reviewed journal, *Science*. It is now available in hard- and soft-cover forms.

Knox, P., and L. McCarthy. *Urbanization*. 2nd ed. Upper Saddle River, NJ: Pearson, 2005. A clear textbook that focuses on the social and demographic trends and impacts of an urbanizing world population.

Krupp, F., and M. Horn. *Earth: The Sequel, The Race to Reinvent Energy and Stop Global Warming*. New York: Norton, 2009. This book is a glimpse at the power of alternative thinking to meet the growing challenge of providing sustainable energy for the world's population.

Kurlansky, M. *Cod: A Biography of the Fish That Changed the World*. New York: Penguin, 2003. A corker of microhistory and ecology that helps connect the human and biotic forces that shape and have been shaped by this extraordinary species.

Leopold, A. *A Sand County Almanac: With Essays on Conservation.* London: Oxford University Press, 1949. This is required reading for anyone with a hint of interest in ecology or conservation. It is a sacred text among many ecologists.

Leopold, L. *A View of the River.* Cambridge, MA: Harvard University Press, 2006. Leopold considers the full nature of the geology and structure of rivers from a systems approach.

Levin, S. *Fragile Dominion: Complexity and the Commons.* Cambridge, MA: Perseus, 1999. A very useful analysis of the systems approach to understanding ecology and the risk of community-level decline and interactions.

Manning, R. *Against the Grain: How Agriculture Has Hijacked Civilization.* New York: North Point Press, 2004. Manning has sounded the alarm about homogenized food production and the risks associated with a reliance on big agriculture.

Marzluf, J., E. Shulenberger, W. Endlicher, M. Alberti, G. Bradley, C. Ryan, C. ZumBrunnen, and U. Simon, eds.. *Urban Ecology: An International Perspective on the Interaction Between Humans and Nature.* New York: Springer, 2008. This volume covers a wide range of issues from across the world in the emerging field of urban ecology.

Masood, E., ed. *Dry: Life without Water.* Cambridge, MA: Harvard University Press, 2006. Masood captures the ecology of life in arid regions. The story of human culture in xeric ecosystems is a potent reminder of the risk we face as water becomes globally scarce.

McDonough, W., and M. Braungart. *Cradle to Cradle: Rethinking the Way we Make Things.* New York: North Point Press, 2002. This volume investigates the industrial process as an outgrowth of our technology and looks for ways to reinvent the life cycle of products so that they can be produced and consumed more sustainably.

McKibben, B. *Deep Economy: The Wealth of Communities and the Durable Future.* New York: Times Books, 2007. McKibben tackles the hard questions that emerge from the tragedy of the commons and the ecology of human-dominated landscapes.

Morton, T. *Ecology without Nature: Rethinking Environmental Aesthetics.* Cambridge, MA: Harvard University Press, 2007. A wonderful critical analysis of the idea of place, which is at the core of our personal environmental narrative.

Pielou, E. *After the Ice Age: The Return of Life to Glaciated North America*. Chicago, IL: University of Chicago Press, 1991. This text is a groundbreaking analysis of the succession of species that occurred in North America following the retreat of the Wisconsin Glacier.

Platt, R., ed. *The Humane Metropolis: People and Nature in the 21st Century*. Amherst: University of Massachusetts Press, 2008. This volume provides multiple views of the biological and social ecology of livable and just urban landscapes.

Reisner, M. *Cadillac Desert: The American West and Its Disappearing Water*. New York: Penguin, 1986. This book and the subsequent PBS film stand as a chronicle of the politics and ecology of the water issues of the American West, especially in the development of Los Angeles.

Quamman, D. *The Song of the Dodo: Island Biogeography in the Age of Extinctions*. New York: Scribner, 1996. Quamman provides a lucid history and analysis of the theory of island biogeography launched by MacArthur and Wilson. The analysis leads to applications in conservation biology.

Scherr, S., and J. McNeely. *Ecoagriculture: Strategies to Feed the World and Save Wild Diversity*. Washington, DC: Island Press, 2002. The authors investigate strategies and sustainable techniques that will reduce the load of anthropogenic nutrients added to our soils. This is a key part of a model for environmental recovery.

Sutherland, W. *From Individual Behavior to Population Ecology*. New York: Oxford University Press, 1996. Sutherland crosses many scales in his analysis of the levels of ecological interactions.

Timothy, G. *Minamata: Pollution and the Struggle for Democracy in Postwar Japan*. Cambridge, MA: Harvard University Press, 2001. This volume provides a riveting account of the ecology and politics of one of the world's worst pollution disasters.

Wackernagel, M., and W. Rees. *Our Ecological Footprint: Reducing Human Impact on Earth*. Gabriola Island, BC: New Society Publishers, 1996. This is a bold attempt to quantify the impact of our daily choices on the Earth's environment. This work has served as a foundation for many ecosystem studies.

Weart, S. *The Discovery of Global Warming*. Cambridge, MA: Harvard University Press, 2008. Weart traces the story of how we came to understand the facts and characteristics of global climate change.

Wilson, E. O. *Consilience: The Unity of Knowledge.* New York: Vintage Books, 1998. A Pulitzer Prize-winning scientist presents a controversial but engaging view of the bringing together of all knowledge into a scientific and ecological framework. Guaranteed to ignite an argument around any dinner table.

———. *The Creation: An Appeal to Save Life on Earth.* New York: Norton, 2006. Written as a letter from a scientist to a pastor, this is a broad and heartfelt plea for the coming together of all cultures in service to a sustainable planet.

———. *Sociobiology: The New Synthesis.* Cambridge, MA: Harvard University Press, 1975. Wilson takes animal behavior and evolution by storm with his game-changing book on the nature of animal societies. This book engaged a whole generation of scholars in a vigorous debate about the animal nature of the human species.

Worldwatch Institute. *The State of the World: Into a Warming World.* New York: Norton, 2009. A fabulous compendium of environmental data that are organized into a coherent but sobering analysis of current world trends.

Wright, R. *Environmental Science: Toward a Sustainable Future.* 10th ed. Upper Saddle River, NJ: Pearson Prentice Hall, 2008. This college text is a great compendium of environmental science at the interface of human sustainability.

Notes

Notes